T0325821

Writing for Science and Engineering

Writing for Science and Engineering

Papers, Presentations and Reports

Second Edition

Heather Silyn-Roberts

Department of Mechanical Engineering
Auckland University
Auckland, New Zealand

AMSTERDAM • BOSTON • HEIDELBERG • LONDON • NEW YORK • OXFORD
PARIS • SAN DIEGO • SAN FRANCISCO • SINGAPORE • SYDNEY • TOKYO

Elsevier
32 Jamestown Road, London NW1 7BY
225 Wyman Street, Waltham, MA 02451, USA

First edition 2000
Second edition 2013

Notices

Knowledge and best practice in this field are constantly changing. As new research and experience broaden our understanding, changes in research methods, professional practices, or medical treatment may become necessary.

Practitioners and researchers must always rely on their own experience and knowledge in evaluating and using any information, methods, compounds, or experiments described herein. In using such information or methods they should be mindful of their own safety and the safety of others, including parties for whom they have a professional responsibility.

To the fullest extent of the law, neither the Publisher nor the authors, contributors, or editors, assume any liability for any injury and/or damage to persons or property as a matter of products liability, negligence or otherwise, or from any use or operation of any methods, products, instructions, or ideas contained in the material herein.

British Library Cataloguing-in-Publication Data
A catalogue record for this book is available from the British Library

Library of Congress Cataloging-in-Publication Data
A catalog record for this book is available from the Library of Congress

ISBN: 978-0-08-098285-4

This book has been manufactured using Print On Demand technology. Each copy is produced to order and is limited to black ink. The online version of this book will show color figures where appropriate.

Contents

Introduction: How to Use This Book

The goal of this book is to be relentlessly practical. It has been designed with the specific needs of science and engineering graduate students and junior professionals in mind. It is the result of working with hundreds of you in Europe and Australasia to learn how you access information, the type of information you want and the sorts of books you don't like. These are the sorts of things that many students have told me:

- **You appreciate prescriptions.** This book certainly gives prescriptive guidelines. It's very user-centred – almost a recipe book. In places, it even gives formulae, e.g. for structuring material in the various sections of a journal paper. It was this feature of the first edition that students liked, and I've expanded on this concept in the second edition. Science and technical writing *can* be guided to a great extent by prescriptions. You may not achieve high style, but you'll get professional competency.
- **You don't like dense text.** I've heard many disparaging comments about books with 'too many words'. A great many of you have said that you don't want to read long passages of text; most of you prefer concise, listed material to long paragraphs.
- **You need to be able to read any chapter in isolation.** In this book, you don't have to have read the previous parts of the book to understand the later ones.
- **Looking things up and cross-referring is second nature to scientists and engineers.** This book has lots of cross-references within it to other parts of the book.
- **You appreciate knowing the mistakes to avoid, and that you are not alone in your difficulties.** This book lists the common difficulties and errors.
- **Many of you have not had enough guidelines on the requirements of technical writing and presentation during your undergraduate years.** This book assumes no basic knowledge, but it is not simplistic.

Added for the Second Edition

Since the publication of the first edition, I've spent 15 years running programmes for many hundreds of graduate students in German and Swiss graduate schools associated with major universities and Max Planck Institutes. All of them are doing cutting-edge science in areas ranging from molecular biology, neurobiology, plasma physics, medicine and microbial genetics to intelligent systems, computational neuroscience, biophysical chemistry and more.

The subject areas are wide-ranging, yet the students have the same types of problems and make similar mistakes. They are also under pressure to publish in high-impact journals.

The twice-yearly experience of intensive small-group teaching has been ideal for updating and developing the concepts in this second edition. My thanks to all these students; the education has flowed in both directions. Any mistakes, of course, are all mine.

The Basic Structure of the Book

Chapter 1	**Structuring a Document: Using the Headings Skeleton**	How to decide on a structure for a document
Chapter 2	**The Core Chapter: Sections and Elements of a Document**	The requirements for all the sections likely to be found in a graduate document.
Chapters 3–14	**Specific types of documents**	The requirements for each type of document. Extensively cross-referred to Chapter 2: *The Core Chapter*.
Chapter 15	**Referencing: text citations and the *List of References***	The conventions for referencing within the text and for the *List of References*.
Chapter 16	**Conventions used in scientific and technical writing**	The conventions for such things as formatting equations, rules for capitalisation, etc.
Chapter 17	**Revising and proofreading**	The techniques for revising a document and proofreading the final version or editor's page-proofs.
Chapter 18	**Problems of style: recognising and correcting them**	Recognising and correcting common problems of writing style.
Chapter 19	**A seminar or conference presentation**	The techniques for a formal oral presentation
Chapter 20	**A presentation to a small group**	The techniques for a presentation to a small panel of people, e.g. PhD oral or a design presentation.
Appendix 1	**SI units**	
Appendix 2	**The parts of speech; tenses and forms of the verb**	
Appendix 3	**Recommended Scientific Style manuals**	

How to Use This Book

If you need the primary information about the following: • **The basic skeleton of headings of a technical document** • **How to choose sections for a document** • **How to guide a reader through a document**	See Chapter 1: *Structuring a Document: Using the Headings Skeleton*
If you need information about how to write a specific section or element of a document	See Chapter 2: *The Core Chapter: Sections and Elements of a Document*
If you need to write a specific type of document	See Chapters 3–14. Go straight to the chapter covering that specific type of document. It will be extensively cross-referred to the detail you need in other parts of the book.
If you need to prepare a seminar or conference presentation	See Chapter 19: *A Seminar or Conference Presentation*
The other chapters give the supporting information on the conventions of technical documentation: referencing, editorial conventions and revising.	Chapters 15–17

Section 1

Document Structure. The Requirements for each Section

1 Structuring a Document: Using the Headings Skeleton

This chapter covers:

- *TAIMRAD*: the classic structure of an experimental report.
- When *TAIMRAD* isn't an appropriate structure for your document.
- The basic skeleton of section headings.
- Building an extended skeleton of section headings.
- Using the *Outline* mode of Microsoft Word® to help organise your document.
- The importance of overview information: building a navigational route through your document.
- Deliberate repetition of information in the basic skeleton.

The Basic Skeleton of Section Headings for a Technical Document

This section covers:

- The classic *TAIMRAD* structure for an experimental report.
- When *TAIMRAD* isn't suitable: choosing section headings.
- The basic set of headings forming the skeleton of a document, whatever its topic or length.

TAIMRAD: The classic structure of an experimental report

The classic, traditional structure for an experimental report, particularly a journal paper, is the *TAIMRAD* structure: *Title*, *Abstract*, *Introduction*, *Methods*, *Results and Discussion.*

When TAIMRAD Isn't an appropriate structure for your document

The classic *TAIMRAD* structure may not be suitable if you are reporting on:

- Experimental work but the structure needs to be expanded from the restrictive *TAIMRAD* form

  or

- Work that is not of an experimental nature.

Writing for Science and Engineering.
DOI: http://dx.doi.org/10.1016/B978-0-08-098285-4.00001-7

In this case, you will need to construct your own set of headings.

There is no single structure that can be applied to all reports. The following sections give guidelines for this.

Choosing a set of main headings

The basic skeleton of all professional technical documents is made up of a set of main headings. The headings don't depend on the topic or length of the report, or whether it presents experimental or investigational work that you've done, or material that you've researched only from the literature (e.g. a generalised project report).

Documents tend to start and end with the same sections; the middle part will depend on the subject matter of your document. To show this, Table 1.1 compares the basic format for a generalised short and a long document.

Table 1.1 A Basic, General Skeleton for a Generalised Short and a Long Document to Show the Similarities

A Short Document	**A More Complex Document** (*Note: You may not need all of these sections*)
Title	**Title page**
Summary	**Abstract** or **Summary** or **Executive Summary**
	Recommendations (if needed)
	Acknowledgements
	Table of Contents
	List of Illustrations
Glossary of Terms and Abbreviations or **List of Symbols** (may not be needed)	**Glossary of Terms and Abbreviations** *or* **List of Symbols**
	Theory (if needed)
Introduction *or* **Background**	**Introduction** *or* **Background**
(Middle part of text – *your choice of headings*)	(Middle part of text – *your choice of headings*)
	Discussion
Conclusions	**Conclusions**
Recommendations (if needed) *Alternatively, placed immediately after the Summary*	**Recommendations** (alternative position) *or merged as*
	Conclusions and Recommendations
	Acknowledgments (alternative position)
	References *and/or* **Bibliography**
Appendices (may not be needed)	**Appendices**

Choosing Section Headings: Building an Extended Skeleton

This section describes how to:

- Build up the general skeleton into an appropriate extended skeleton of sections for your document. This covers every type of document that is not of a strictly *TAIMRAD* structure.
- Use the standard sections frequently used in longer documents.

Steps to take

Step 1: Working from the basic skeleton, plan an enlarged skeleton for your document. Use Table 1.2 for help: it does the following:

- It lists many standard sections used in postgraduate science and technological documents in the approximate order in which they would occur in the document.
- It gives the purpose of each section.
- It cross-refers you to the pages of this book that give guidelines on how to write each section described in Table 1.2.

Step 2: Work out your own headings for the central part of the document. Think about what the reader needs.

- Ask yourself: ***What does the reader need to be able to assess my material most readily? How can I best tell this story for the reader?***
- Don't ask: *How do I want to present this material?* This is quite different; it is looking at it from your point of view, not the reader's. Documents that are written from the writer's point of view run the risk of being difficult for a reader to readily understand.

Table 1.2 To Determine What Sections You Will Need for a Document

Section Heading	Purpose of Section	Frequency of Use	Cross Reference (Unless otherwise stated, the material is in Chapter 2 – The Core Chapter: Sections and Elements of a Document.)
Title	To adequately describe the contents of your document in the fewest possible words	Necessary	See **Title**, page 19
Title page	This is usually the covering page (first page) of a document, giving the title of your document, information about yourself and your institution, and any declaration that you may need to make	Longer documents	See **Title Page**, page 21

(*Continued*)

Table 1.2 To Determine What Sections You Will Need for a Document (Continued)

Section Heading	**Purpose of Section**	**Frequency of Use**	**Cross Reference (Unless otherwise stated, the material is in Chapter 2 – The Core Chapter: Sections and Elements of a Document.)**
Abstract *or* **Summary** *or* **Executive Summary**	• To give readers a miniaturised version of the document, so that they can identify the basic content quickly and accurately • To give readers a *brief* overview of all of the key information. Vitally important to help the readers assess the information in the rest of the document • To help readers decide whether they need to read the whole document	Necessary	See Chapter 3 – **Abstract, Summary, Executive Summary**, page 53 For abstracts in a journal paper, see Chapter 6 – **A Journal Paper**, page 83
Keywords	A brief list of keywords relevant to your document that will be used by electronic indexing and abstracting services	Usually only for a journal paper	See **Keywords**, page 22
Acknowledgements	To thank the people who have given you help in your work and in the preparation of your document	If needed	See **Acknowledgements**, page 23
Table of Contents	Gives the overall structure of the document. Lists the headings and subheadings, together with their corresponding page numbers	Longer documents	See **Table of Contents**, page 23
List of Illustrations	To give a listing – separate from the *Table of Contents* – of the numbers, titles and corresponding page numbers of all your figures and tables	Longer documents	See **List of Illustrations**, page 26

(*Continued*)

Table 1.2 To Determine What Sections You Will Need for a Document (Continued)

Section Heading	Purpose of Section	Frequency of Use	Cross Reference (Unless otherwise stated, the material is in Chapter 2 – The Core Chapter: Sections and Elements of a Document.)
Glossary of Terms and Abbreviations (*or* **List of Symbols)**	To define the specialist terms and abbreviations (including acronyms) that you use in the main text of the document	If needed	See **Glossary of Terms and Abbreviations**, page 27
Introduction	• To allow readers to understand the background to the study without needing to consult the literature themselves. You should keep your reader adequately informed but not write an over-long *Introduction* • To point out the relationships between the various authors' works, - the correlations and contradictions • To show gaps in the knowledge, correlations, contradictions and ambiguities • Having pointed out the gaps in the knowledge, to state the main objective of the work described in your paper (often unclear or missing) • To provide a context for the later discussion of the results • To define specialist terms used in the paper • In a longer document, to describe the structure of the document	Common	See **Introduction**, page 28

(*Continued*)

Table 1.2 To Determine What Sections You Will Need for a Document (Continued)

Section Heading	**Purpose of Section**	**Frequency of Use**	**Cross Reference (Unless otherwise stated, the material is in Chapter 2 – The Core Chapter: Sections and Elements of a Document.)**
Background	Sometimes used as an alternative heading to *Introduction* But where a document needs both an *Introduction* and a *Background* ***Introduction***: usually a restatement of the brief and a description of the structure of the document ***Background***: gives the history of the subject matter and the objectives of the study. Alternatively, the objectives can be stated in a separate *Objectives* section	If needed	See **Background,** page 30
Objectives	To describe the aims of your study	If clear statement needed	See **Objectives,** page 30
Purpose Statement	To state the aims of the document (equivalent of the *Objectives* section)	These four sections are sometimes found in management reports	See **Purpose Statement**, page 31
Scoping Statement *or* **Scope**	To describe the topics covered in the document		See **Scope statement**, page 32
Procedure Statement	To describe the processes you followed in investigating the topic of the document		See **Procedure statement**, page 32
Problem Statement	To describe the problem and its significance		See **Problem statement**, page 32
Literature Review	To review the literature in your field of work. Shows that you have a good understanding of the historical development and current state of your topic	Research document	See Chapter 4: ***A Literature Review***, page 63
A section covering your planning of tasks (suggested headings*: Schedule of Tasks or Time Management*)	To describe how you propose to schedule the various tasks that you will have to do	Often in management reports	See **Schedule of Tasks** ***or*** **Time Management**, page 33

(Continued)

Table 1.2 To Determine What Sections You Will Need for a Document (Continued)

Section Heading	**Purpose of Section**	**Frequency of Use**	**Cross Reference (Unless otherwise stated, the material is in Chapter 2 – The Core Chapter: Sections and Elements of a Document.)**
Allocation of Responsibilities	To describe the person(s) who will be responsible for each task	May be needed in a report from a project team	See **Allocation of Responsibilities**, page 33
Ownership/ Confidentiality	An agreement between you and the commercial organisation funding you that gives you some right of publication of your results, while assuring the organisation that you will not divulge commercially sensitive information	May be needed in a research proposal	See Chapter 5: ***A Research Proposal***, page 35
Requirements	To describe what you expect to need from your funding organisation	If needed	See **Requirements**, page 35
Costs	To describe the expected costs that you are asking the funding organisation to cover	If needed	See **Costs**, page 36
Methods *or* **Materials and Methods** *or* **Procedure**	To describe your experimental procedures. Aim: repeatability by another competent scientist	Research reports	See **Methods**, page 36
Results	To present your results but not to discuss them	Research reports	See **Results**, page 37
Discussion	To show the relationships among the observed facts that you have presented in your document and their significance and to draw conclusions	Common	See **Discussion**, page 38
Conclusions	To present your conclusions, soundly based on the previous material in the document	Necessary	See **Conclusions**, page 39
Recommendations	To propose a series of recommendations for action	If needed	See **Recommendations**, page 40

(*Continued*)

Table 1.2 To Determine What Sections You Will Need for a Document (Continued)

Section Heading	**Purpose of Section**	**Frequency of Use**	**Cross Reference (Unless otherwise stated, the material is in Chapter 2 – The Core Chapter: Sections and Elements of a Document.)**
Suggestions for Future Research	To propose directions for further development of your work	If needed	See **Suggestions for future work**, page 41
References *or* **List of References**	A list of the works that you have cited in the text. Strict conventions govern this process	If your sources have been cited in the text	For full details of the conventions for citing references in the text and compiling the *List of References*, see Chapter 15 – ***Referencing: Text Citations and the List of References***
Bibliography	A list of works that you have not cited in the text, which you think will be of interest to the reader	If your sources have *not* been cited in the text	See **Bibliography**, Chapter 15 – ***Referencing: Text Citations and the List of References***, page 171
Appendices	At the end of a document, complex material that would interrupt the flow of your document if it were to be inserted into the main body. For example: raw data, detailed illustrations of equipment, coding, specifications, product descriptions, charts and so on.	When complex supporting material needed	See **Appendices**, page 42
Index	To provide at the end of a long document, a list in alphabetical order of topics mentioned in the book and the pages where they occur	Longer documents	See **Index**, page 44

It shows (i) the possible standard sections of a graduate technical document, (ii) the purpose of each section, (iii) how often the section is used and (iv) the relevant cross-reference to guidelines in this book.

The *Outline* Mode of Microsoft Word®: Organizing a Document

This section very briefly describes the *Outline* mode of Microsoft *Word*®. This mode helps organise a document, revise it, and produce a professional-looking document.

The *Outline* mode of Microsoft *Word*® will:

1. **Help organise a set of headings and subheadings of various levels.**
   This is useful for the first stage of organising a document. You decide on your headings, the sub-headings and their divisions, and then assign them to their various levels (level 1 for a main heading, level 2 for a sub-heading and so on). They can be easily reassigned to different levels at any time in the writing process.
   The text is then inserted under the headings to produce the full document.
2. **Collapse the document to display only selected levels of headings.**
   This gives an overview of the whole document. You can select the level of overview. By collapsing the document and selecting to display only the level 1 headings, you can check the overall structure of the document in terms of only its main headings. By progressively displaying greater levels of sub-headings, you can obtain an increasingly more detailed view of the structure of the document.
   This also helps in revising the first draft of the document.
3. **Enable a heading to be dragged and dropped to a different place in the document or to a different level.**
   This helps to organise and revise the document. When a heading is dragged and dropped, the corresponding text is also moved.
4. **Automatically produce a *Table of Contents* with the corresponding page numbers.**

The Importance of Overview Information

This section describes how to help readers to navigate their way through your document. In this way, they will understand and assess the information much more readily.

This is done by using the basic skeleton and section summaries to provide overview information.

Even though technical documents have side-headings, they are very often difficult to assess and extract information from. This can be because the readers can't see a route through it, i.e. something to help them to navigate their way.

You can construct a navigational guide through your document by using the basic skeleton and building on it. Your readers should then be able to use this – probably unconsciously – to gain a much readier understanding of your material.

Building a roadmap by giving overview information throughout

Psychological studies have shown that our brains need initial overviews to better assess the full information that follows.

To use this concept, think of structuring information in the shape of a diamond (Figure 1.1).

1. First, **think of the whole document as being diamond shaped.** At the narrow ends, the information is brief, focused and concise.
   - The *Summary* or *Abstract* at the beginning and the *Conclusions* at the end each give overview information.
   - The *Summary* prepares the reader for the whole document; the *Conclusions* confirms the findings and their significance.
2. Next, **think of each section of a long document as also being diamond shaped.** It will have a title; It will also help the readers if it has a very brief summary immediately under the section's title.

This structure does two things:

1. **It helps the expert reader get an undetailed understanding of the key information in a document. The formula for this is as follows: Read only the *Title, Summary/Abstract/ Executive Summary, Conclusions* and *Recommendations.***
   These sections – together with the section summaries – should form a road map that orientates the readers and guides them through the document. They also give the non-expert

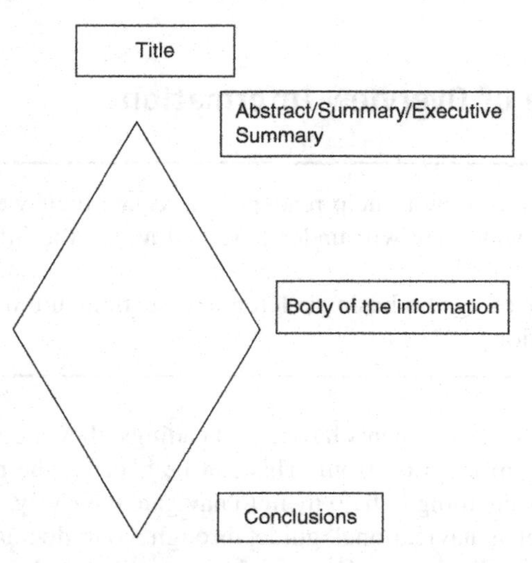

Figure 1.1 Diagrammatic representation of the structure of a complete document. The level of detail is low at the two narrow ends – the initial and final overview information (*Title, Abstract* or *Summary* or *Executive Summary*, and *Conclusions*).

reader a means of obtaining an undetailed overview. (For an explanation of the varying levels of detail delivered by certain sections, see Table 1.3.)

2. **It lets non-expert readers obtain an overview of the document by reading these particular sections, while avoiding the detail.**

How a reader with less expertise than you would probably read a document structured in this way:

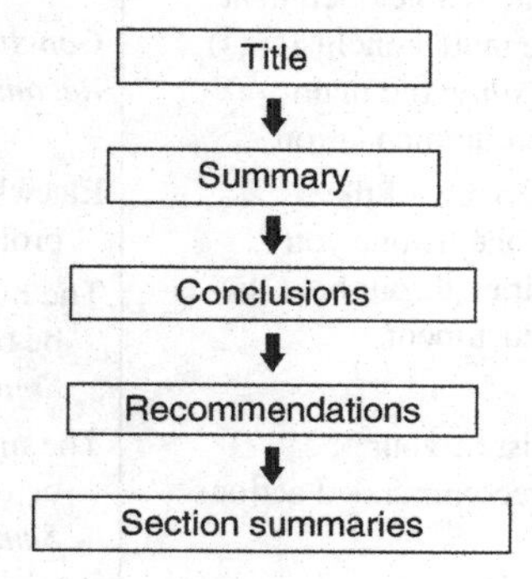

Suggested wording (placed immediately before or after the main Summary):

For overview information about this document, please read the *Summary, Conclusions and Recommendations* together with the section summaries at the beginning of each section.

Table 1.3 Explanation of How the Sections of the Basic Skeleton Deliver Overview Information at Increasing Levels of Detail

Section	**Level of Detail: Increases from First to Fourth**	**What the Section Does for the Reader**
Title	First level	Gives immediate access to the subject matter
Summary or Abstract	Second level	Gives an undetailed overview of the whole document
Conclusions *and* Recommendations	Third level	Gives succinct overview information about your conclusions and recommendations
Each Section Section Summary	Fourth level	*For the middle sections (the sections where you choose appropriate section headings)* Gives overviews of the material in each section If the material in the section does not lend itself to being summarised, substitute a *Scope Statement* that describes the topics covered in the section

Table 1.4 The Deliberate Repetition of Material Throughout a Document

The Section of the Basic Skeleton	The Information	The Places in the Rest of the Document Where the Information is Repeated
Abstract or **Summary** or **Executive Summary**	Undetailed overview of the whole document	Throughout the document
	The main conclusion(s)	*Conclusions*
	Possibly: the main recommendation	*Recommendations*
Conclusions	Overview of the conclusions you draw throughout the document	Elsewhere in the document; probably in the *Discussion* The main conclusion(s) will also be repeated in the *Abstract* or *Summary* or *Executive Summary*
Recommendations	A list of your recommended actions	The main recommendation(s) might be repeated in the *Abstract* or *Summary* or *Executive Summary*
Appendices		Summaries of the *Appendix* material might appear in the main body of the document

Deliberate Repetition of Information in a Document

This section describes how information is deliberately repeated in the various sections of the basic skeleton.

People are sometimes concerned because they see information repeated throughout a report. Remember, however, that this repetition is deliberate and controlled – the basic skeleton calls for it. The repeated information forms part of the navigational route described previously and guides the reader through the document. Table 1.4 shows the information that is repeated and the sections where it occurs.

This deliberate restatement of undetailed information in the basic skeleton is a feature of a professional document. But information that is repeated because the document has been sloppily assembled is another matter.

Specific Types of Documents: Using This Book

This section describes how to use this book if you are writing a specific type of document.

Table 1.5 Specific Types of Documents Dealt with in this Book and the Relevant Chapter and Page Numbers

Type of Document	Relevant Chapter and Page Number
Abstract *or* **Summary** *or* **Executive Summary**: • A short abstract/summary (200–300 words) • A journal paper abstract • A conference abstract (about two pages) • An Executive Summary (10–25% of the whole document)	Chapter 3, page 53
Literature review	Chapter 4, page 63
Research proposal	Chapter 5, page 75
Journal paper	Chapter 6, page 83
Progress report	Chapter 7, page 111
Consulting or Management report **A project team's progress reports** **A recommendation report**	Chapter 8, page 117
Engineering design report	Chapter 9, page 121
Formal letters	Chapter 10, page 125
Emails and faxes	Chapter 11, page 135
Procedure or set of instructions	Chapter 12, page 137
Thesis	Chapter 13, page 143
Conference poster	Chapter 14, page 155
Additional material:	
SI units	Appendix 1: page 257
The parts of speech; forms of the verb	Appendix 2: page 261
Recommended scientific style manuals	Appendix 3: page 265

Specific types of documents are dealt with in Chapters 3–14. Each of these chapters gives extra material relevant to the type of document (including a suggested structure) and is cross-referred to the material in Chapter 2 – *The Core Chapter: Sections and Elements of a Document*. Table 1.5 lists the various specific types of documents covered in these chapters and additional appendix material that may be helpful.

Checklist for the structuring of a document

- ☐ Are you using the necessary headings of the basic skeleton?
- ☐ Are the headings of your expanded skeleton appropriate to your topic?
- ☐ Are your headings in a logical order?
- ☐ Have you built a navigational route for the reader by giving overview information throughout your document: an *Abstract* or *Summary* or *Executive Summary, Recommendations* and *Conclusions*, and in a long report, section summaries?
- ☐ Have you deliberately controlled the repetition of information throughout the document?

2 The Core Chapter: Sections and Elements of a Document

This chapter covers the requirements for each of the sections and elements required in the various types of documents you may have to prepare as a postgraduate. Any one document will not need all of the sections described in this chapter.

Many sections are described under most or all of the following headings:

- **Purpose**
- **Difficulties**
- **How to Write It**
- **Common Mistakes**
- **The Tense of the Verb**
- **Checklist**

For a specific type of document, use the relevant chapter for that document type in combination with this chapter.

The sections are listed in the approximate order in which they are usually found in a document.

The following sections and elements of a document are covered in this chapter. Some elements are cross-referenced to other chapters that give a detailed treatment of that element.

Title of Section or Element	Location of More Specific Material
Letter of Transmittal or Covering Letter	Chapter 10: *A Formal Letter (Hardcopy or Online)*, page 125
Title	*Title*, Chapter 6: *A Journal Paper*, page 19
Running Title	*Running Title*, Chapter 6: *A Journal Paper*, page 20
Title Page	
Authorship and Affiliation (particularly in a journal paper)	*Authorship and Affiliation*, Chapter 6: *A Journal Paper*, page 87
Abstract/Summary/Executive Summary	Chapter 3: *Abstract, Summary, Executive Summary*, page 53 *Abstract*, Chapter 6: *A Journal Paper*, page 88
Keywords	*Keywords*, Chapter 6: *A Journal Paper*, page 87

Writing for Science and Engineering.
DOI: http://dx.doi.org/10.1016/B978-0-08-098285-4.00002-9

Title of Section or Element	Location of More Specific Material
Acknowledgements	*Acknowledgements*, Chapter 6: *A Journal Paper*, page 102
Table of Contents	
List of Illustrations	
Glossary of Terms and Abbreviations *or* List of Symbols	
Introduction	*Introduction*, Chapter 6: *A Journal Paper*, page 91
Background	
Theory	
Objectives	
Purpose Statement	
Scope Statement	
Procedure Statement	
Problem Statement	
Literature Review	Chapter 4: *A Literature Review*, page 63
Schedule of Tasks *or* Time Management (*or similar*)	
Allocation of Responsibilities	
Ownership/Confidentiality	
Requirements	
Costs	
Materials and Methods	*Materials and Methods*, Chapter 6: *A Journal Paper*, page 93
Results	*Results*, Chapter 6: *A Journal Paper*, page 96
Discussion	*Discussion*, Chapter 6: *A Journal Paper*, page 98
Conclusions	*Conclusions*, Chapter 6: *A Journal Paper*, page 101
Recommendations	
Suggestions for Future Research	
References *or* List of References	Chapter 15: *Referencing*, page 169
List of Personal Communications	Chapter 15: *Referencing*, page 175
Bibliography	Chapter 15: *Referencing*, page 86
Appendices	
Index	
Illustrations: Figures and Tables	

Many of these sections are described under most or all of the following headings:

1. **Purpose** or the aim of each of the sections.
2. **Difficulties** when writing the section.
3. **How to write** the section.
4. **Common mistakes** to avoid.
5. **The tense of the verb to use.**
   See Appendix 2 for guidelines for using tense in technical documents, as well as definitions and examples of the various tenses of the verb.
6. **Checklist.**

Letter of Transmittal, Covering letter

Letters that accompany a document. See Chapter 10: *Formal Letters (Hardcopy or Online)*, pages 32.

The Title

For a journal paper title – see *Title*, Chapter 6: A Journal Paper, page 86.

Purpose

- To give the reader immediate access to the main point or subject matter.
- To give informative, not generalised, information.
- To describe the contents of paper in the fewest possible words. Note: This can lead to a clumsy string of nouns. See *Noun Trains*, Chapter 207, page 214.

Difficulties

Devising a title that is:

- Short enough
- Contains all of the key information
- Makes sense (i.e. is not ambiguous, does not contain problematic syntax)

How to write it

For the rules of capitalisation in a title, see *Title case*, Chapter 16: *Conventions*, page 196.

- Work out the information that a reader would need to gain immediate access to the main point or subject matter of your document.
- It should not be too general, not be too detailed and should contain the necessary key information.
- After your efforts to make it short, make sure that it makes sense. The structure can be lost during the quest for the minimum number of words, making it muddled and ambiguous.
- If the title is a long string of words, it may be effective to rewrite it into two parts using a colon. See *A Hanging Title*: Chapter 6: *A Journal Paper*, page 86.

Running Title

The short title required by journals for the tops of the pages. See Chapter 6: *A Journal Paper*, page 87.

A Conference Poster Title

See also *Planning the Poster*, Chapter 14: *A Conference Poster*, page 155.

- **A poster title needs to contain the key information but also to draw the attention of the poster viewers.** For this reason, it can be shorter, more punchy and possibly more querying or controversial than a title for a journal paper. Questions as titles can be provocative but they can also imply that your results are in question. If the title is too long, it will take up too much space and overwhelm the poster since the letters will be large.

  Long, informative and suitable for a journal paper.

  Dropping rates of elaiosome-bearing seeds during transportation by ants (*Formica polyctena* Foerst.): implications for distance dispersal.

  Gives the conclusion; shorter; more direct.

  Dispersal distance of elaiosome-bearing seeds is determined by ants' dropping rates.

  A question attracts the viewers' attention, but it could imply that your results are ambiguous.

  Do ants' dropping rates determine the distance dispersal of elaiosome-bearing seeds?

  Too short to contain the required information but possibly suitable for a poster.

  Do ants affect the dispersal of seeds?

- People at conferences can be interested in new methodology. If you have used a novel method, show it in the title.

  A new method for detoxification of mycotoxin-contaminated food.

- Place the title at the top of your poster. Don't be tempted to try a trendy configuration such as placing it vertically up one side.

Common mistakes

1. Uninformative. Too general or too colloquial; no good indication of the content.
2. Too long and clumsy.
3. Misleading or ambiguous; does not accurately reflect the content.

Checklist for the *Title*

- ☐ Does it give the reader immediate access to the main point or subject matter of your work?
- ☐ Does it use the fewest possible words and still make sense?
- ☐ Is it too long?
- ☐ Is it too general?
- ☐ Is it too detailed?
- ☐ Does it make sense?

The Title Page

This is usually the covering page (first page) of a longer document.

Purpose

To give the title and information about yourself and your institution and any declaration that you may need to make.

How to write it

In general, it should state:

- The title of your document.
- Your name and department, university or institution. (For guidelines on multiple authorship, see *Authorship and Affiliation*, page 87, Chapter 6: *A Journal Paper*)
- The date of submission.
- The name of the relevant person, organisation or tertiary level course to which it is being submitted.
- ***Other possible elements:***
  - A declaration that it is your own work may be needed.
    Typical wording:

    I declare that this report is my own unaided work and was not copied from or written in collaboration with any other person.
    Signed…
  - ***For a thesis***: The degree for which the thesis is being submitted and the institution.
    Typical wording:

    A dissertation submitted in partial fulfilment of the requirements for the degree of Doctor of Philosophy in the University of Middletown
- **Layout**
  - Graduate courses will often have specific instructions about how to lay out the title page.
  - It should make a pleasing arrangement, not crowded, with plenty of empty space.
  - It should be free of gimmicks such as *ClipArt* pictures.

Checklist for the *Title page*

Does the Title Page show the following:

☐ An informative title
☐ Your name
☐ The name of your department/faculty/organisation
☐ The date of submission
☐ *Possibly*: a declaration that it is your own work
☐ *For a thesis*: the degree for which the thesis is being submitted and the name of the institution to which the thesis is being submitted?

Authorship and Affiliation (Particularly in a Journal Paper)

See *Authorship and Affiliation*, Chapter 6: A Journal Paper, page 87.

Abstract (can also be called a Summary)

The *Abstract* or *Summary* is important for the understanding of the whole document. It is often poorly written and does not give adequate information.

For a journal paper *Abstract*, see *Abstract*, Chapter 6: *A Journal Paper*, page 87.

For overview information in documents other than a journal paper, see Chapter 3: *An Abstract, a Summary, an Executive Summary*, page 53, which gives information on the following:

- **The different types of content in a *Summary* or *Abstract***: descriptive, informative and descriptive/informative.
- **The short type of *Summary* or *Abstract*** that is part of a larger document. It is generally from 200 words to half a page but will be longer in a thesis or large document.
- **An *Abstract* for a conference paper** (usually about two pages).
- **An Executive Summary**.

Keywords

These are needed only in a journal paper. See *Keywords*, Chapter 6: *A Journal Paper, page* 87.

Acknowledgements

Purpose

To thank staff and other people who have helped by:

- Sending you material (experimental or literature)
- Giving you technical help in your laboratory work
- Discussing your work with you
- Putting you in touch with other people
- Giving you emotional support, particularly members of your family

How to write it

- State very simply that you would like to thank the following people and state also the type of help they gave you.
  Example:
  I would like to thank the following people:
- If you feel particularly grateful to someone, start by saying ***I am particularly grateful to****… for… I would also like to thank…, and* then list their names and state what they did.
- Make sure that you include the following:
  - Not only the surname of a person you are thanking but also the first name or the initials.
  - The person's correct title (Dr, Associate Professor, Ms, Mr and so on). If you don't know it, make a point of finding it out, if necessary by telephoning their institution.
  - Their department/institution/organisation.

Common mistakes

- Using flippant wording. It is possible to sound patronising or silly.
- Not including people's first names or initials and their department and institution. Wrong: *I would particularly like to thank Dr Stevens for giving me samples of…* Corrected: ***I would particularly like to thank Dr A. J. Stevens, Department of Evolutionary Biology, University of Middletown, for giving me samples of…***

Table of Contents (*or* Contents Page)

Purpose

To give a listing of the headings and subheadings, together with their corresponding page numbers.

Difficulties

- Deciding on an appropriate layout.
- Formatting it correctly so that the indentations are consistent.
- Making sure that the page numbers in the text correspond with those on the Table of Contents.

All of these problems can be eliminated by using the facility on your word processor that automatically constructs and formats your *Table of Contents*. See ***The Outline Mode of Microsoft Word®: Organising a Document***, Chapter 1: *Structuring a Document: Using the Headings Skeleton*, page 11.

How to write it

If you are not using the word processor's facility, use the following guidelines:

- Decide the lowest level of heading to display on the *Table of Contents* (e.g. whether you want to go down to the subheading level or to the sub-subheading level).
- List all of the sections and all of their subheadings down to your chosen level down the left-hand side of the page.
    - Number the sections and their subheadings by the accepted conventions, using the decimal point numbering system (see *Numbering of Illustrations, Sections, Pages, Appendices, Equations*, Chapter 16: *Conventions*, page 192).
    - If you are indenting for subheadings, make sure that the indentations are consistent for each level of heading.
- Place the corresponding page numbers at the right-hand side of the page.
    - Use the accepted conventions for the page numbering system (see *Numbering of Illustrations, Sections, Pages, Appendices, Equations*, Chapter 16: *Conventions*, page 192).
- Don't list individual figures on the *Table of Contents*. If you have a lot of illustrations (e.g. as in a thesis), you need a section called List of Illustrations, which immediately follows the *Table of Contents* (*see* List of Illustrations, *this chapter*).
- Conventionally, the *Abstract* or *Summary* is not listed on the *Table of Contents*. However, it may help the reader to do so, even though it is placed immediately after the *Title page* and is therefore easily found. This has been done in the example below.

Common mistakes

1. Mismatches between the text and *Table of Contents* in the wording and numbering of the various headings, together with their corresponding page numbers.
2. Inconsistent formatting and indenting of the various levels of headings.

Example: Table of Contents

Table of Contents

Checklist for the Table of Contents

To use while checking these features: see *Numbering of Illustrations, Sections, Pages, Appendices, Equations*, Chapter 16: *Conventions, page* 192.

- ☐ Does it list the preliminary pages and give their page numbers in Roman numerals system?
- ☐ Does it list the following:
  - Chapter headings
  - Section and subsection headings
  - The List of References section
  - Each appendix?
- ☐ Does it give the correct section number of the sections, subsections, the *List of References* section and each appendix?
- ☐ Does each appendix have a title?
- ☐ Do the page numbers match up with those in the text?
- ☐ Is it consistently formatted? Are the indentations of the sections and subsections consistent?
  Note: You can avoid the last two problems by using the automatic Table of Contents function in your word-processing software.

List of Illustrations

Purpose

To list – separate from the *Table of Contents* – the numbers, titles and corresponding page numbers of all your tables and figures.

How to write it

- The term *illustrations* includes tables and figures (graphs, line drawings, photographs, maps and so on). Use the title *List of Illustrations* if your document contains both tables and figures. If it contains only tables, call it *List of Tables*; if only figures, use *List of Figures*.
- If you are using *List of Illustrations*, list all of the figures first, followed by a list of all of the tables.
- List the number, title and page of each illustration.
- Place the *List of Illustrations* immediately after the *Table of Contents*. If both of them are brief, put them on the same page with the *Table of Contents* first.

Common mistakes

Mismatches often occur between the features of the text figures and tables and the way they are listed in the *List of Illustrations*. Use the automatic Table of Contents function in your word-processing software to avoid this.

Checklist for the List of Illustrations

To use while checking these features: see *Numbering of Illustrations, Sections, Pages, Appendices, Equations*, Chapter 16: *Conventions*, *page* 192.

- ☐ Are all of the figures listed first and then the tables?
- ☐ Are the number, title and page of each illustration given?
- ☐ Do the page numbers match up with those in the text?
- ☐ Conversely: **In the main text**, do the illustration numbers in the text match those in the *List of Illustrations*?

Glossary of Terms and Abbreviations (or List of Symbols, when dealing with only mathematical symbols)

Purpose

To define the specialist terms, symbols and abbreviations (including acronyms) that you use in the main text of the document.

How to write it

- Decide terms that need definitions. Remember that a term self-evident to you may not be as generally known as you think. Even when you are writing a specialist document that will be read only by experts – such as a thesis – your referees will appreciate a list of clearly defined terms.
  Make sure, though, that you don't include terms that are generally very well known; to define them would look silly.
- Terms that need to be dealt with include the following:
  - Specific technical terms.
  - Greek or other symbols.
  - Abbreviations (usually called acronyms). These are often in the form of the initial letters in capitals of a series of words, e.g. **PCR: polymerase chain reaction; PLC: programmable logic controller.**
    List the terms in alphabetical order of the abbreviations, followed by the definition of each one, e.g.:
    PCR **polymerase chain reaction**
    PLC **programmable logic controller**
- Before you list the terms and abbreviations, it may be appropriate to state the following:
  S.I. (Système International d'Unités) abbreviations for units and standard notations for chemical elements, formulae and chemical abbreviations are used in this work. Other abbreviations are listed below.

Where to put it

The *Glossary of Terms* can be placed either at the beginning of the document immediately after the *Table of Contents* or the *List of Illustrations* (this is the optimal position for the reader), or at the end, immediately before the Appendices.

If the glossary is large, and you feel that it needs to be at the end of the document, readers would appreciate a note placed immediately before the *Introduction*, referring the readers to the page number of the glossary. Suggested wording:

Explanations of terms and abbreviations used in this document are given in the *Glossary of Terms and Abbreviations*, page 27.

Introduction

For material specific to a journal paper *Introduction*, see Chapter 6: *A Journal Paper*, *page* 91.

Purpose

1. To allow readers to understand the background of the study without needing to consult the literature themselves. You should keep your reader adequately informed but not write an overly long *Introduction*.
2. To point out the relationships between the various authors' works – the correlations and contradictions.
3. To show gaps in the knowledge, correlations, contradictions and ambiguities.
4. Having pointed out the gaps in the knowledge, to state the main objective of the work described in your paper (often unclear or missing).
5. To provide a context for the later discussion of the results.
6. To define specialist terms used in the paper.
7. In a longer document: to describe the structure of the document.

Note: For guidelines for writing a self-standing literature review, see Chapter 4: *A Literature Review*, page 63. This is often required as a chapter in a thesis or as a separate assignment.

Difficulties

The *Introduction* can sometimes be a difficult section to write due to the following most common problems:

- **Writing too much**. It can happen you waste time including too much material in to an *Introduction* that then needs to be cut down.
- **Deciding how much background detail to include.** This is especially difficult when your readers are made up of both specialists and non-specialists.
  For specialists: a thorough introduction to the topic may sound patronising.
  For the less knowledgeable: too little information may leave them unclear about what you are trying to achieve. Graduate students trying to gain familiarity with the subject often find *Introductions* too short and uninformative.

- **Deciding how many references to include.**
- **Writing a good first sentence.** The first sentence shouldn't be a banal statement of general knowledge. It needs to provide an overall introduction, but be specific to your particular problem. It can be difficult to think up, and people sometimes resort to a trite statement of the obvious. For example:

  Toxic waste is a very serious problem in the world today.

  Even pompously dressing it up can't disguise the banality:

  The quantity of toxic waste currently generated in the world is a problem of the utmost seriousness.

- **Long documents: final paragraphs:** After stating your purpose and approach, in a fresh set of paragraphs, briefly describe the structure of the document.
  Section 4 gives the historical background…Section 5 reviews the current techniques… and so on.
- **For a document that contains both an *Introduction* and a *Literature Review*:** The *Introduction* in this case will be made up of a description of the general background to the study, with only a few references, together with a description of the structure of the document.

Tense of the verb in the Introduction

See *The Correct Form of the Verb*, page 224, Chapter 18, *Problems of Style*, for guidelines on using tense in technical documents, together with examples of the various forms of the tenses of the verb.

The *Introduction* needs a mixture of present and past tenses.

Example:
It has been previously shown (*past*) that plants flower (*present, because it's established knowledge*) under environmental conditions that maximise seed set and development… Much work has been done (*past*) towards understanding the environmental, physiological and genetic regulation of flowering in the species under study… *Author name* (20xx) showed that the mutants flowered (*past*) later than wild-type plants; GI was therefore proposed (*past*) to be a floral promotion gene. This work describes (*present*) research undertaken to verify the isolation of…

Common mistakes

- Too long, rambling, unspecific, unstructured, with irrelevant material.
- Specialist terms not defined.

Checklist for the Introduction

For a checklist for a journal paper *Introduction*, see Chapter 6: *A Journal Paper*, page 93.

Does the Introduction do the following:

- ☐ Adequately reviews other people's work?
- ☐ Identifies the gaps in the knowledge and inconsistencies in this area of research?
- ☐ Gives a historical account of the area's development? (if appropriate)
- ☐ Puts your study into the context of other people's work?
- ☐ *Clearly* states the purpose of your study in the final paragraphs?
- ☐ Follows the purpose statement by briefly summarising your approach?
- ☐ Briefly describes the structure of the document? (*in a long document*)

Background

If you are writing a report, you may prefer to call your section *Background* instead of *Introduction*. If so, all of the guidelines given previously for an *Introduction* apply also to a *Background* section.

Some organisations, in particular consulting engineers, may require both an *Introduction* and a *Background*. In this case, the differences are the following:

- The *Introduction* usually includes a restatement of the brief and a description of the structure of the report.
- The *Background* gives the history of the subject matter and the objectives of the study. Alternatively, the objectives can be stated in a separate *Objectives* section.

Theory

If a description is needed of the theoretical background of your work, it should be written for a busy professional in your discipline who has a good, broad understanding of the area but no detailed knowledge. This means that the description of the theory should not start at an elementary level and should include the material that you think such a person would need and be no longer than needed.

Objectives

Purpose

To describe the aims of your study.

How to write it

- This section should be very brief and concisely stated.
- The objectives can be listed:

  The objectives of this study were:

  1. **To (*establish the…*)**
  2. **To (*determine the…*)**

- In a longer document such as a thesis or a research proposal, it is effective to first state the broad purpose and then follow with the specific objectives:

  Aims of this study
  The purpose of this study was to investigate the development and structure of bacterial biofilms grown on different specific substrata in a subsurface-flow wetland.

  The specific objectives of the research were to:
  - **develop methods for investigating biofilms grown on different substrata in a constructed wastewater treatment wetland**
  - **investigate the initial adsorption of bacteria to different wetland substrata, namely…**
  - **study the early development of biofilms and their population structures.**

Purpose Statement, Scoping Statement (or Scope), Procedure Statement, Problem Statement

Some documents, particularly consulting or management documents, require one or more of these four sections. They present material that in other reports is covered in sections such as the Introduction, Background, or Procedure/Methods. They answer the following questions:

Question	*Section that Answers the Question*
What is the purpose (objective) of this report?	Purpose statement
What is the problem? What is its significance? Who or what caused the writer to write about the problem?	Problem statement
What are the specific topics and their limits that the report covers?	Scope statement
What procedures were used to investigate the problem?	Procedure statement

Purpose statement

Purpose

To state the purpose of the document (the equivalent of the *Objectives* section).

How to Write It

1. State the purpose clearly. 'The purpose of this report is…' (to solve whatever problem made the report necessary or to make whatever recommendation).
2. Name the alternatives if necessary.

Problem statement

Purpose

To describe the problem and its significance

How to write it

Probably in this order:

- Describe the problem, giving the basic facts about it.
- Explain what has gone wrong.
- Specify the causes or the origin of the problem.
- Describe the significance of the problem (short term and long term).
- Give the appropriate data and state their sources.
- Specify who is involved and in what capacity.
- Discuss who initiated the action on the problem or what caused you to write the report.

Scope statement

Purpose

To describe the topics covered in a report.

How to write it

- **For a feasibility study or recommendation report:** name the criteria you used to formulate the requirements.
- **For other types of reports:** identify the main sections or topics of the report.
- Specify the boundaries or limits of your investigation.

Procedure statement

Purpose

To describe the processes you followed in investigating the topic of the report. This statement establishes your credibility by showing that you took all the proper steps. In a standard experimental report, this material is covered in the *Procedure/Materials and Methods/Methodology* section.

How to write it

Explain all of the actions you took, including the people you interviewed, research performed and so on.

Literature Review

Purpose

- To review the literature in your field of work.
- To show that you have a good understanding of the history and current state of your topic.

See Chapter 4: *A Literature Review*, page 63.

A Section Covering Your Planning of Tasks

Suggested headings for this section are **Schedule of Tasks *or* Time Management.**

Purpose

To describe how you propose to schedule the various tasks that you will have to do.

How to write it

This involves intelligent and informed guesswork. The most convenient way of showing a time schedule is to use a Gantt chart. This subdivides your proposal into tasks together with the dates when you propose to begin and end each one. It is a version of a bar chart (see Figure 2.1).

Points to remember when compiling a Gantt chart:

1. You need to assess:
   - The number of tasks
   - How long each task is going to take
   - How you can fit each task with another
2. **Most tasks will overlap with each other**. For instance, if your first task is to get a preliminary understanding of the literature, this is likely to overlap with the first stage of your experimental process.

Allocation of Responsibilities

This section may be needed in a report from a project team.

- In a preliminary report, describe the person(s) who will be responsible for each task and the roles of the subsidiary individuals.
- In a final report, you may need to give a more detailed account of the various roles that each person has played in the progress of the work and the writing of the report. In addition, you may be asked for a peer review of each person. This calls for an objective assessment of the effectiveness with which each person fulfilled his or her roles.

PLANNING SCHEDULE – 20xx

Research Project: **Modification of a pulsatile pump for an isolated heart**

No.	*Activity*	*Feb*	*Mar*	*Apr*	*May*	*June*	*Estimated no. of hours*
1.	Literature search						20
2	Test and assess the current pumping system						40
3	Design the modifications to the current system						20
4	Build components						40
5	Assemble, test and evaluate the rig						60
6	Modify and retest the rig if necessary						50
7	Evaluation of the final results						40
8	Write the final report						60
						Total Time (hours)	330

Figure 2.1 Example of a Gantt chart for time scheduling.

Ownership/Confidentiality

Purpose

An agreement between you and the commercial organisation funding you that gives you some right of publication of your results, while assuring the organisation that you will not divulge commercially sensitive information.

When needed

This section may be needed in a research proposal.

For academic projects, it is essential that the rights to publish scientific papers and theses are retained. On the other hand, many commercial organisations will want to own the rights to the outcomes of your research so that they can commercialise them. Moreover, all commercial organisations will expect you to maintain in confidence any commercially sensitive information that they provide to help the research or that results from the project.

How to write it

The wording of the proposal should be tailored to the specific circumstances. Terms for ownership should be agreed to before any work starts, but at the proposal stage, it may be sufficient to state that these are to be negotiated.

Requirements

Purpose

To describe what you expect to need from your funding organisation during your research.

How to write it

Possible wording:

To complete this work, we will need the following from (*name of the organisation*):

Then give a list of your requirements. The types of things that you may have to request are:

- Guaranteed access to the field site, the organisation's laboratories, test hall and so on.
- Access to specified items of the organisation's equipment.
- Assistance with the preparation and/or installation of specified items of equipment.
- A laboratory base with access to power, water and bench space.

Costs

Purpose

To describe your expected costs during the course of your research that you are asking the funding organisation to cover.

How to write it

It needs to an itemised list of the various costs. You may need to include such things as the following:

- Student stipend
- Supervisory costs
- Materials and equipment
- Travel costs
- Overheads

State the total final cost (the so-called *bottom line*).

Materials and Methods (can also be called Methods or Procedure)

> See also *Materials and Methods*, Chapter 6: *A Journal Paper*, page 93, for detailed material on how to write it, tense of the verb, common mistakes and a checklist. This material is also useful for any document that describes experimental work.

Purpose

- To describe your experimental procedures.
- To give enough detail for a competent worker to repeat your work.
- To describe your experimental design.
- To enable readers to judge the validity of your results in the context of the methods you used.

Difficulties

Not many. This section is often the easiest part of a document to write. Describing experimental methods is usually very straightforward.

Therefore it is often the best place to start writing. Writing a document is often difficult, and there is absolutely no need to write it in sequence from beginning to end. Start with the section that will give you the fewest problems.

See also *Starting to Write a Journal Paper*, Chapter 6: *A Journal Paper*, page 84.

Results

See also *Results*, Chapter 6: *A Journal Paper*, page 96, for detailed material on how to write it, tense of the verb, common mistakes and a checklist. This material is also useful for any document that describes experimental work.

Purpose

- To present your results, but not to discuss them.
- To give readers enough data to draw their own conclusions about the meaning of your work.

Difficulties

Deciding how much detail to include.

General comments

- Your results need to be clearly and simply stated. This is the new knowledge that you are presenting to the world; it is the core section of the document.
- **It needs to be presented as a logical story.** If it is interrupted by material that is too detailed or is not directly relevant, your readers are going to become disorientated and lose the thread.
- It is often the first place that readers familiar with the topic will look (after the *Title* and the *Abstract*). Many readers, after first reading the *Abstract*, then look at the illustrations. (This highlights the need for illustrations to be as self-explanatory as possible, by means of informative titles and captions.)

Structuring of Corresponding Headings for *Materials and Methods* and *Results* Sections

If your work has a number of separate experimental elements to it, group the procedures and the results for each element together. Don't describe all of the procedures in sequence, and then follow with all the results in sequence. This will result in a poorly structured document that is difficult for your assessor to read.

If appropriate, you can include a short *Discussion* section for each separate part and follow up with an overall main *Discussion*.

Schematic

Efficient Structure	*Poor Structure*
First experiment Procedure Results Discussion **Second experiment** Procedure Results Discussion **etc.** • • • **Overall Discussion**	**Procedures** Description for first experiment Description for second experiment Description for third experiment Description for fourth experiment **Results** Results for first experiment Results for second experiment Results for third experiment Results for fourth experiment **Discussion**

Results and Discussion

If it is possible to write a *Results and Discussion* section, then do so; it is often easier to write and better for the reader than separate *Results* and *Discussion* sections.

Formula for a *results and discussion* section

Efficient Structure
First set of results Discussion of them Second set of results Discussion of them etc. • • • **Overall *Discussion* incorporating the conclusions**

Discussion

See also *Discussion*, Chapter 6: *A Journal Paper*, page 98 for detailed material on how to write it, tense of the verb, common mistakes and a checklist. This material is also useful for any document that describes experimental work.

This section is almost always required in a journal paper and is often appropriate to other types of documents.

In a *Discussion* section, show the relationships among your observations and place them into the context of other documentation and other people's observations and work.

Purpose

1. To show the significance of your results and how your results lead to your conclusions.
2. To explain how your results support the answer.
3. To show the relationships among your observations.
4. To put the results into context.
5. To state clear conclusions, unless there is a *Conclusions* section in your document.

Difficulties

Not knowing where to start, what or how much to put in it and how to create a logical flow.

Conclusions

See also *Conclusions/Conclusion*, Chapter 6: *A Journal Paper*, page 101 for detailed material on how to write it, tense of the verb, common mistakes and a checklist. This material is also useful for any document that describes experimental work.

Purpose

To present the conclusions that arise from the material in the document.

How to write it

- *Important*: There should be no new material in this section. Each conclusion must be drawn directly from material that has already been presented in the main body of the report, and it must be well substantiated.
- Each conclusion should be related to specific material.
- Each conclusion should be brief (because the full explanation is given elsewhere in the document).
- A numbered or bulleted list can be used if appropriate. Start with the main conclusion and then present the conclusion points in descending order.
- If there are a large number of conclusions, they can be grouped under headings as shown in the box. In this case, ensure that the numbers run sequentially through the whole list to make them more readily identifiable.

Example: *Conclusions* section

Public health issues

1. **The waters of the basin continue to present a significant risk to human health owing to sewage pollution.**
2. **Sewer overflows are a major source of pollution; they contribute over 98% of the faecal coliform pollution load to the basin and half of the nitrogen load.**
3. **etc.**

Water quality

4. **The water in the basin is very highly loaded with nutrients, mainly from sewage overflows.**
5. **Storm water is also a major source of pollution.**
6. **etc.**

Siltation

7. **Sedimentation plates indicate that the level of siltation in the basin has increased from 3 to 6.5 mm per year over the past 20 years.**
8. **Sediment quality is fair and is not considered to present an abnormal health risk.**
9. **etc.**

Recommendations

Purpose

To propose a series of recommendations for action, resulting from the conclusions drawn from your work (e.g. a design improvement, management strategies and so on).

Position in the document

In a formal technical document, the section *Recommendations* is usually placed either.

- At the start of the document, immediately after the Summary
- At the end of the document, often usefully combined with the *Conclusions* into a section called *Conclusions and Recommendations.*

How to write it

1. Recommendations are your subjective opinions about the required course of action, but this doesn't mean you should go into wild flights of fancy.
2. **Recommendations can be of various types.** Their character will depend on the purpose of your report, e.g.:
   - To choose a new procedure or technique and show why it is preferable
   - To identify a need and suggest a way to fill it

- To explore a new concept and show how it should be applied to existing problems
- To propose a new project and show why and how it should be carried out
- To analyse a problem, find a solution and propose a remedy

3. Recommendations are usually best given as a numbered list. Each item should be brief.
4. Make the main solution to the problem your first recommendation. This usually fulfils the purpose of the report.
5. Then list your other recommendations in a logical way.
6. No recommendation should be unsupported. The supporting information should exist elsewhere in the document.

Tense of the verb

See Appendix 2 for guidelines for using tense in technical documents, along with definitions and examples of the various tenses of the verb.

The conditional, subjunctive or present form of the verb should be used. It can also be worded as a series of instructions (the imperative form of the verb).

Examples:

It is recommended that:

The test equipment should be modified as shown in Figure 4.3 (*conditional*).
The test equipment be modified as shown in Figure 4.3 (*subjunctive*).
The test equipment is modified as shown in Figure 4.3 (*present*).

or:

The recommendations are:

Modify the test equipment as shown in Figure 4.3 (*imperative*).

Checklist for the *recommendations*

- ☐ Is the first recommendation in your list the most important one?
- ☐ Are the other recommendations presented in descending order of importance?
- ☐ Is each recommendation brief, clearly stated and unambiguous?
- ☐ Is each recommendation feasible?
- ☐ Is each one related logically to material presented elsewhere in the report?

Suggestions for Future Research

Purpose

To propose directions for further development of your work.

Research usually opens up more questions than you have time to answer. Many people are unwilling to draw a line under their research and start writing up because it always seems that just a bit more work will tie it up better.

This can be professionally acknowledged by including this section. It may be called *Suggestions for Future Development* or *Suggestions for Further Study*. Check with your supervisor; for reasons of competitiveness, some may be unwilling to indicate future directions.

How to write it

Outline the following:

- Suggestions for the immediate development of the work, i.e. the areas you may not have been able to tie up to your satisfaction. If your project is to be developed by a subsequent student, this will be very useful to both student and supervisor.
- The possible long-term development of the work, if any.

List of References *or* References

See Chapter 15: Referencing, page 176.

List of Personal Communications

See Chapter 15: *Referencing, page* 175.

Bibliography

See Chapter 15: *Referencing*, page 186.

Appendices

> Note the singular and plural of the word, which is sometimes a source of confusion: one *Appendix*; two or more *Appendices*.

Purpose

The appendices are for complex material that would interrupt the flow of your document if it were to be inserted into the main body, e.g. raw data, detailed illustrations of equipment, software coding, specifications, product descriptions, charts and so on.

How to assemble the Appendices

1. Appendices are placed at the end of the document.
2. Material included in an appendix should be there for a specific purpose. It is all too easy to use the Appendices as a sort of rubbish bin into which you tip all the bits you've collected and don't know what else to do with. To avoid an irrelevant, jumbled mess, be selective.

3. Appendices should contain well-structured information, not a formless mass.
4. Related material should be grouped into separate Appendices.
5. Give each Appendix a number or a letter (e.g. Appendix 1, Appendix 2,... *or* Appendix A, Appendix B,...). See *Numbering of Appendices*, Chapter 16: *Conventions*, page 194.
6. Give each appendix a title following the appendix number.
   Example – **Appendix 3: Input File Formats.**
7. The number and title of each appendix should be listed in the *Table of Contents* (*see Table of Contents*, this chapter, page 23).
8. Every item that is included in an appendix should be referred to at an appropriate place in the text.

For Twintex TPP specifications, see Appendix 3: *Technical Data and Specifications*, page 13–21.

Common mistakes

1. Too much unrelated and unnecessary material in the appendices.
2. Lack of organisation of the material.
3. Lack of numbering and titling of each appendix.
4. Some or all of the *Appendices* not referred to in the text.

Checklist for the *Appendices*

To use while checking these features, see *Numbering of Illustrations, Sections, Pages, Appendices, Equations*, Chapter 16: *Conventions*, *page* 194.

☐ Is the body of the document unnecessarily cluttered? Could some of the material be more appropriately placed in the *Appendices*?
☐ Is there material in the *Appendices* that might be better placed in the main text of the thesis? Does it weaken the argument to have it in the *Appendices*?
☐ Do the *Appendices* look like a rag-bag of assorted bits gathered together because you didn't know what else to do with them?
- If so, can you order the material more logically?
- Does it all really need to be included?

☐ Are complex sets of data in the *Appendices* summarised at the appropriate points in the main body text?
☐ Is each *Appendix* titled?
☐ Does the title of each *Appendix* appear in the *Table of Contents*?
☐ Is each *Appendix* referred to in the text?
☐ Is the first reference to it in the text at the first appropriate point, or should it be earlier?
☐ Should you refer to it again later in the text?
☐ Have you given enough details for your examiner to be able to interpret the appendix?
☐ Are the *Appendices* numbered or lettered consecutively?
☐ Does the title of each *Appendix* correspond with that listed in the *Table of Contents*?

☐ Does each illustration in the *Appendices* have the following:
- An appropriate number (e.g. Figure 1: Appendix B)
- An informative title
- Enough information to be interpreted

☐ Is each page in the *Appendix* numbered?

Index

Purpose

At the end of long documents, a list in alphabetical order provides topics mentioned in the book and the pages where they occur. An index makes your material more readily accessible to your readers.

How to compile an index: guidelines

- A helpful index is compiled with the readers' needs in mind.
- Think about how you would look for this item in an index, i.e. how you would classify it, what subsections you would look under and to which other entries you might cross-refer.
- There are features available in word-processing software that allow you to mark the items that should be included in an index. However, a manually assembled index that takes into account the search terms a reader might use is often much more intelligent.
- Include every important subject, topic, subtopic and proper name.
- Entries are usually not capitalised.
- Most indexes consist of two levels of entries: main headings and subheadings, if necessary. Occasionally, a third-level heading may be needed.
- Use cross-references at appropriate places to guide the reader to other related entries in the index. A cross-reference is usually placed either after the main heading or at the end of the list of subheadings, with *see also* in italics before the cross-reference.

Example of a three-level index entry with a cross reference to another entry:

hovering, 218–234 ***see also*** **flight**
  actuator disc theory, 218–226
  blade element theory, 226–230
    applied to animals, 230–232
  hummingbirds, 224–235
  mandarin fish, 232–233
  wasps, 233–234

Illustrations

This section gives general guidelines about the subject of illustrations in general. It does not aim to tell you how to produce effective graphics. Here your best guide is the expertise of other students. There is also a wealth of information online that can be helpful.

Definition of terms

Caption/legend

Each illustration has a figure number, a title and an explanation. This explanation is supplied by the *caption* and the *legend*. These two terms are sometimes distinguished from each other, but they are more often confused or considered synonymous (*A Manual of Style*, University of Chicago Press, 1969). To avoid confusion, this book uses only the term *caption* to describe the explanatory material that follows the title of an illustration.

For the conventions for numbering illustrations, see *Numbering of Illustrations*, Chapter 16: *Conventions, page* 192.

Each illustration must have a clear purpose. Ask yourself what your readers will need to help understand the text. For instance, they may not need a diagram of a standard piece of equipment, but they may appreciate a diagrammatic representation of how you modified it. Figures therefore need to be as self-explanatory as possible.

Make each illustration as self-contained as possible. Remember, to make a preliminary assessment of the work, readers often skim through a document looking first at the figures before they read the text. Each graph, table or diagram should not need the reader to refer to the text to make its overall meaning clear.

Guidelines for illustrations:

1. Make sure each illustration has the following:
   - **An informative title** (*for the principles of devising titles, see page* 19, *The Title, this chapter*).
   - A **clear, explanatory caption** following the title.
2. **Use as few abbreviations as possible in an illustration. If you must use abbreviations, include a key in the illustration itself.** For instance, if you have been sampling at sites that you have, for convenience, called AO3, BV4 and so on, try not to use these as headings in a table. Instead, do one of the following:
   - Think up short labels that give a better description of the characteristics of each site.
   - Include a key to the labels in the illustration.
3. **Make your illustrations look professional.** Find out from other students the best software packages to use.
4. **Emphasise the data, not the axes.** The axes should be thinner than the curve or plot lines. Most graphing programs will do this.
5. **Make sure that the axes on each graph are fully labelled.** Each axis must be labelled with what is being plotted and the units. (***Common mistake:*** leaving out the units.)
6. **Don't extend the axes too far.** The *X* and *Y* axes should extend only to the next tick mark after the maximum value for the data. Graphing programs usually do this.
7. **Make sure that there are not too many lines on your graphs.** It is better to create two graphs than to have one that is overcrowded. Graphs with more than four lines are likely to be difficult to read, especially if the lines overlap.

8. **Make each line on a graph easily distinguishable from the others.** Some software packages do not create symbols that adequately distinguish one line from another.
9. **Include in an appendix the raw data on which important graphs are based.** Refer to the appendix at the appropriate point in the text. This enables the reader to assess the exact values of the data.
10. **Make sure that your illustrations will be big enough.** It is very easy to compress an illustration until it is almost unreadable. Take particular care that any labelling is large enough, particularly subscripts and superscripts.
11. **Keys:** Usually the key is given in the illustration itself, but sometimes it can be included in the title.
12. **Numbering of illustrations:** To use while checking these features, see *Numbering of Illustrations, Sections, Pages, Appendices, Equations*, Chapter 16: *Conventions, page* 192.
    With all the reformatting that a large document may need, it is all too easy to get the figures out of sequence or to refer in the text to a figure that doesn't exist. It is worthwhile to use the feature of a word-processing package that automatically correlates text references to a figure with its figure number.
13. **If your document has a lot of illustrations:** You need to include a *List of Illustrations*: for the conventions; see *List of Illustrations*, page 26, this chapter.
14. **Using other people's illustrations and data:** Cite the source at the end of the caption to the illustration, and include the source in your *List of References*. See *Copying and Adapting Illustrations*, Chapter 15: *Referencing*, page 175.

Checklist for *Figures*

To use while checking these features, see *Numbering of Illustrations, Sections, Pages, Appendices, Equations*, Chapter 16: *Conventions*, page 192.

- ☐ Is the figure needed?
- ☐ Could it be simplified?
- ☐ If it is a graph, are there too many lines? Would it be better to consider having more than one graph to illustrate the point?
- ☐ Is the material better presented as an illustration in the text or as an illustration in the *Appendices*?
- ☐ Is each figure numbered consecutively, logically and consistently?
- ☐ Is there enough detail in the figure's title, caption (legend) and keys for an overall interpretation of the figure without reference to the text?
- ☐ Does the title correspond with that given in the *List of Figures*?
- ☐ If you have used or modified someone else's figure, or used someone else's data to construct your own figure, have you done the following:
  - Cited the source in the caption (legend) to your figure?
  - Used the wording required by referencing conventions?
  - Cited the source in your *List of References*?
- ☐ *For any document other than a journal paper*: Is the figure close to but following the place where it is first mentioned in the text?

Checklist for *graphs*

- ☐ Are the x and y axes labelled?
- ☐ Are the units of measurement stated on the axes?
- ☐ Are the lines clearly distinguishable from each other?
- ☐ Are the symbols marking the points clearly distinguishable from each other?
- ☐ Are all the components of the figure clearly labelled?
- ☐ Are abbreviations explained? If not, are they well known? Are you sure?
- ☐ Does the arrangement of the figure proceed from left to right?
- ☐ Does the figure look cluttered and illogical?
- ☐ Is the figure correctly positioned on the page?
- ☐ Is the raw data of important graphs presented in the *Appendices*?

Designing Tables

- For more information about *Tables* for posters, see *Tables* in Chapter 14: *A Conference Poster*, page 161.

Good design of tables can help the reader find the information efficiently. Disorganised, poorly designed or cluttered tables are visually hurtful and can easily put a reader off.

When to use a table

- When you do not need to show trends pictorially (such as a graph).
- When you need to present accurate data and specific facts (e.g. graphs with data that have to be interpolated and are therefore only approximate).
- To demonstrate the relationships between numerical and/or descriptive data.

Do not use unnecessary tables. Your data might be better presented as a graph or given more concisely in the text.

Guidelines for designing tables

The Parts of a Table

- **Boxhead:** The horizontal region across the top of the table containing column headings.
- **Stubhead:** The vertical column to the far left of the table in which you list the various line headings that identify the horizontal rows of data in the body of the table.
- **Spanner head:** A region that spans the head of two or more columns. Used for related parameters and to reduce repetition in the column heads.
- **Body spanner:** A region that spans across two or more columns in the body of the table.
- **Column heads:** Must all have headings as described here:
  - The headings should include units of measure, where appropriate, and any scaling factors used.
  - Headings should be short. A maximum of two lines is a general rule. If absolutely necessary, use abbreviations and define them in footnotes. But avoid abbreviations if at all possible (Figure 2.2).

<table>
<tr><td rowspan="2">Stub head</td><td rowspan="2">Column head</td><td rowspan="2">Column head</td><td colspan="2">Spanner head</td><td rowspan="2">Box Head</td></tr>
<tr><td>Column head</td><td>Column head</td></tr>
<tr><td></td><td colspan="4">Body spanner</td><td></td></tr>
<tr><td></td><td>Column entry</td><td></td><td></td><td></td><td></td></tr>
<tr><td></td><td colspan="4">Body spanner</td><td></td></tr>
<tr><td></td><td>Column entry</td><td></td><td></td><td></td><td></td></tr>
</table>

Figure 2.2 The parts of a table.

Direction of reading information

- Information always reads *down* from the boxhead.
- Information reads *down* from the stubhead.
- Information described by the stubhead reads *across*.

Guidelines for designing tables

- The independent variable (e.g. time) usually reads across the table.
- The dependent variable (e.g. test number) reads vertically.
- Every column or spanner head needs a unit of measurement (or some explanation if the values are arbitrary rather than measurements).
- It is more clear to put the unit in the head rather than in the entries, e.g.:

This is better...	**...than this**
Temperature (°C)	*Temperature*
40	40°C
60	60°C
100	100°C

- **Spanner heads** help to combine data and avoid repetition. Instead of repeating the unit of measurement after two or more column heads, a spanner head can be used:

Average daytime temperatures °C	
1999	2000

- **Body spanners** are effective ways to divide data sets, e.g. data from men, data from women.
- **Important:** Columns are easier to compare than rows because we are more used to running our eyes down a column to compare data than running our eyes across. In Table 2.1,

Table 2.1 Maximum Tensile and Compressive Stresses of the Bones of the Wing of *Pteranodon ingens*

Bone	Tensile Stress σ_t (MPa)	Compressive Stress σ_c (MPa)
Metacarpal	68	73
First phalanx	70	75
Second phalanx	113	123
Third phalanx	215	230
Fourth phalanx	146	174

From M. Johnston (1997) An aeroelastic model for the analysis of membrane wings and its application to yacht sails and *Pteranodon ingens*

Table 2.2 Maximum Tensile and Compressive Stresses of the Bones of the Wing of *Pteranodon ingens*

Bone	Metacarpal	First Phalanx	Second Phalanx	Third Phalanx	Fourth Phalanx
Tensile stress σ_t (MPa)	68	70	113	215	146
Compressive stress σ_c (MPa)	73	75	123	230	174

Note: This is the same data as in Table 2.1, but the rows and columns are reversed.

it is very easy to see by running down the columns that the third phalanx of *Pteranodon ingens* has higher tensile and compressive stresses than the other bones. This is not obvious from Table 2.2, where the data in the rows and columns have been transposed.

- Plan your table so that there is adequate spacing of columns and to avoid splitting the table across two pages.
- Any table too wide to fit upright on a page should be presented in landscape mode so that it is read from the right-hand side of the page.
- **The table number and title are placed *above* a table; the title of a figure is placed *below* it.** This is just one of those strange conventions that you have to stick to.
- **Make the table as self-contained as possible.** Readers will often look at tables and figures first to assess the key points of the results without first reading the text. To make sure that it contains the key information needed, do the following:
  - Give it an **informative title**. The readers should be able to understand the table without looking for the relevant part of the text. The title should typically include these items:
    - The independent variable(s)
    - The dependent variable(s)
    - The concept or the species studies
- A **comprehensive but concise caption (legend)** that includes definitions of the symbols.
- **Footnotes** can be used for the following:
  - To define abbreviations
  - To explain a missing entry

- To explain an entry that seems anomalous
- To explain where an entry had different conditions from those in the rest of the table
- To expand a shortened entry

  But don't let footnotes take over the table. If there are too many, you need to reassess the method of presentation of the data. A table may not be appropriate.

- **If any table is taken from another source**, the reference should be cited in the legend to the table. See *Copying and Adapting Illustrations*, Chapter 15: *Referencing*, page 175.

Checklist for *Tables*

To use while checking these features, see *Numbering of Illustrations, Sections, Pages, Appendices, Equations*, Chapter 16: *Conventions*, page 192.

- ☐ Is the table needed?
- ☐ Could the data be better presented as a figure?
- ☐ Is the material better presented as a table in the text or as a table in the *Appendices*?
- ☐ Does each table deal with a specific question?
- ☐ Does the table have a clear, uncluttered layout? Could it be simplified?
- ☐ Does each table show what the text says it shows?
- ☐ Are all the tables numbered consecutively?
- ☐ Is your numbering system consistent throughout the document?
- ☐ Does the title correspond with that given in the *List of Tables* or *List of Illustrations*?
- ☐ Does the page number where the table appears correspond to that given in the *List of Tables* or *List of Illustrations*?
- ☐ Are all of the tables of a similar format?
- ☐ Does each table have an informative, explanatory title?
- ☐ Is the wording of the stubhead and the boxhead(s) also contained in the title of the table?
- ☐ Is there enough detail in the table's title, caption (legend) and keys to interpret the table?
- ☐ Is the table as self-explanatory as possible, without the reader having to refer to the text to understand it?
- ☐ Are symbols and abbreviations explained? If not, are they well known? Are you sure?
- ☐ Are you using too many decimal points?
- ☐ Are there missing or extra numbers?
- ☐ Are units of measurement stated?
- ☐ Is there too much detail?
- ☐ Are column entries aligned?
- ☐ Are column headings short (no longer than two lines)?
- ☐ Is the table positioned correctly on the page?
- ☐ *For a document other than a journal paper*: Is the table close to but following the place where it is first mentioned in the text?
- ☐ *For a journal paper*: Does it match the journal style (as set out in the *Instructions to Authors*) in *all* of its features?

Section 2

Specific Types of Documents

3 An Abstract, a Summary, an Executive Summary

Note: For detailed information about a journal paper *Abstract*, see *Abstract*, Chapter 6: *A Journal Paper*, page 88.

This chapter covers:

- The purpose of an *Abstract/Summary/Executive Summary*
- Definitions: *Abstract/Summary/Executive Summary*
- Difficulties in writing
- General information for all types of *Abstracts and Summaries*
- The different types of content of an *Abstract/Summary* (descriptive, informative and descriptive/informative)
- Length of an *Abstract or Summary*
- A conference paper *Abstract* (two to three pages)
- An *Executive Summary*: purpose, length and format
- Common mistakes of *Abstracts* and *Summaries*
- Checklists

Purpose of an Abstract/Summary/Executive Summary

- To give readers a miniaturised version of the document, so they can identify the key information quickly and accurately.
- To provide a navigational tool for the whole document. Overview information is very important in helping the reader understand and assess the information in the rest of the document.
- To help readers decide whether they need to read the whole article.
- To help conference organisers decide from your **conference abstract** whether to invite you to present a paper.

Writing for Science and Engineering.
DOI: http://dx.doi.org/10.1016/B978-0-08-098285-4.00003-0

Definitions: Abstract/Summary/Executive Summary

> Note: In technical documentation, the words *Abstract* and *Summary* are often used interchangeably to mean the same thing. The specific differences are given below.

In different circumstances, you may be asked to write an *Abstract*, a *Summary* or an *Executive Summary*. Each of these presents an overview of the material in your document, but they differ in their purpose and wording.

An *Abstract* presents the overview to an expert audience. *Required* in specialised documents such as journal papers, conference papers and posters.

A *Summary* presents the overview to a less-specialised audience. Anyone reading it should be able to gain an understanding of the main features and findings of your document, without the detail. *Required* at the beginning of every document you write, if an *Abstract* or *Executive Summary* is not specifically asked for.

An *Executive Summary* presents the overview to an executive audience in non-specialist language as far as possible. It is generally longer than a *Summary* and should present the work in greater detail than a standard *Summary* does. *Required* in a management or consulting document, specifically for the management personnel of an organisation. They may have no scientific or technical expertise. The language therefore needs to be understood by non-experts.

Difficulties

Many people find it extremely difficult to adequately summarise their document in a specified number of words. The difficulties arise from:

- Not knowing the difference between descriptive and informative abstracts.
- Deciding on the core information.
- Making sure that all aspects are covered.
- Making sure that the abstract is not concentrating on only some aspects at the expense of others.
- Linking the information into a coherent story.
- The final cutting-down process. For instance: you have written a good 400-word abstract, but only 300 words are stipulated. The final process of paring it down even further, without dropping important information, can often be very difficult.

How to Write It: General Information for All Types of Abstract or Summary

1. **The elements of information needed – in this order – are the following:**
   a. **A statement that places your work in context**
      This is a statement that presents the big picture. But avoid an overall statement of generally known fact.

b. Your method of investigating it (*if appropriate*)
This might be a description of an experimental technique, an analytical method, a design technique, a system design and so on.

c. Your main results or observations
This could be an experimental finding, a theoretical result, an improved design or system, a body of information (if you have done a literature search on a specific topic) and so on.

d. Your main conclusion(s)
Your deduction about what your work means. An *Abstract/Summary* should contain only one or two main conclusions; the complete set of conclusions is then presented in a *Conclusions* section (see *Conclusions*, Chapter 2: *The Core Chapter*, page 39).

e. Your main recommendation(s) (*if appropriate*)
If you have several recommendations, use a section called *Recommendations* (see *Recommendations*, Chapter 2: *The Core Chapter*, page 40).

2. **Aim for an informative, not a descriptive summary/abstract** (see below). This is important.
3. **An *Abstract/Summary* should not contain any information that does not appear in the main body of the document.**
4. **Don't use tables, figures or literature references in a brief *Abstract/Summary*.** However, they are needed in a conference abstract of the longer type (*see A Conference Abstract*, this chapter), and they can be used in a long *Executive Summary* if appropriate.
5. **Write the final version of the *Abstract/Summary* after you have completed the paper.** If you write it early in the process to focus your thoughts, revise it later. It is vital to get the same emphasis and perspective as is in the main body of the paper.
6. **Section summaries.** Larger documents may benefit from having short summaries at the beginning of each section, in addition to the main *Abstract/Summary*. Section summaries give an overview of the information in that section and are useful navigational tools for the reader. Each one should be headed *Section Summary*. (See Chapter 1: *Structuring a Document: Using the Headings Skeleton*, page 13.)

Aiming for an Informative Abstract/Summary

Based on their content, *Abstracts/Summaries* are generally classified into the following types:

1. **Descriptive** or **indicative.** This type should be avoided.
2. **Informative**. This is the type to aim for.
3. **Informative–descriptive**. A thesis may need this type.

1. The Descriptive or Indicative *Abstract/Summary*

Example of a descriptive *Abstract/Summary*. Avoid writing this type.
Title of document: On-road monitoring of ambient carbon monoxide levels

Abstract
This study aims to measure the on-road spatial distribution of levels of carbon monoxide, a health hazard known to be increasing in Middletown. Methods of measurement are discussed, and the difference between on-road and fixed-site

data is analysed. The influence of temperature, wind speed and humidity is considered. Conclusions as to the effectiveness of this method of carbon monoxide monitoring are given, together with suggested recommendations for future air-quality sampling programmes.

You need to actively avoid writing this type. This describes the structure of the document. It does not give the main findings and conclusions but instead acts more as a *Table of Contents.*

This structural description is generally used only in a long, self-contained literature review. Post-graduate writing nearly always needs the informative type of abstract (see below). Assessors, journal editors and conference organisers can specifically ask for the descriptive type to be avoided. Care is needed to avoid drifting into its typical phrasing and structure.

How to recognise this type of *Abstract/Summary*. Many people write this type in the mistaken belief that this is what is needed. You can recognise it by the following:

- **It describes the *structure* of the document, instead of giving the facts.**
- **It gives no *real* information.** It doesn't help the reader understand what the writer actually did and concluded.
- **It uses stock phrases that are easily recognised.** If you find yourself writing any one of the following words or phrases, you can be almost sure that you are in a sentence that describes structure instead of one that gives real information:

...is analysed/analyses
...is considered/considers
...is described/describes
...is discussed/discusses
...is examined/examines
...is presented/presents
...is given/gives

2. **The Informative *Abstract/Summary***

For an example of an informative abstract, the following is the rewritten version of the descriptive *Abstract* above.

Title of document: **On-road monitoring of ambient carbon monoxide levels**

A statement to place your investigation in context To state why you have done the study (the gap in the knowledge)	This study measures the on-road distribution of levels of carbon monoxide, a health hazard known to be increasing in Middletown and compares the levels with those obtained from fixed-site monitors. Data from fixed sites have been previously used in air-quality monitoring programmes, but there has been doubt about their accuracy in determining levels of carbon monoxide at the adjacent on-road sites.

Method of investigation The results, quantitatively expressed	Levels of carbon monoxide at 1.5 m above road level were monitored during commuter traffic at peak hours, using a moving vehicle on a selected route where fixed monitors were located. The on-road concentrations were found to be greater by three times than those recorded at the adjacent fixed sites (mean values of 11.4±2 SD ppm and 3.9±1 SD ppm relatively). Levels were also found to increase with decreased temperature and wind speed, and increased relative humidity.
Your conclusions	It is concluded that fixed-site data are significantly under-representing ambient levels, and that the methods were effective in measuring the spatial distribution of carbon monoxide, estimating commuter exposure and assessing the effectiveness of fixed site monitors.
Your recommendation	An on-road monitoring programme is recommended as a supplement to the present system of monitoring air quality.

Note: Do not refer to any figures or cite any references in a Summary.

Aim for this type of abstract. It describes the purpose of the work, the methods, results, the main conclusion(s) and possibly the main recommendation(s) as briefly and quantitatively as possible. It is almost 100% certain that this is the type of abstract needed for any post-graduate writing.

- **For an experimental investigation**: It gives specific, quantitative information about methods, results and conclusions.
- **For other types of document**: It gives specific information about the topic under investigation, including hard facts and your main conclusions.
- **Wording:** It avoids the stock phrases of a descriptive *Abstract/Summary* (see above).
- **If you find yourself using one of these stock descriptive phrases when you are writing an informative *Abstract/Summary***, it probably means that a piece of information seems too large to summarise. Reassess it and work out the information that the reader needs.

3. The informative–descriptive abstract

This is a combination of the two types, which gives specific information about the main results, together with general information about the contents of the rest of the document.

A Ph.D. or master's thesis could possibly need this type of abstract. The final results should be specifically stated, and the various kinds of supporting information should be outlined in a more general way. But you need to make sure that you don't slip into too many generalisations.

Length of an Abstract or Summary

For a conference or journal, there will be a stated word number requirement.

When it is your choice: it should be brief: a lengthy abstract defeats its purpose. As a rough guide to lengths:

- A short document (up to 2000 words): 200–300 words may be enough.
- A relatively long document: 300 words to half a page.
- A Ph.D. thesis: usually about 500–800 words or one to two pages.

Abstract specifically for a journal paper, see *Abstract*, Chapter 6: *A Journal Paper*, page 83.

This material gives detail about structure and signalling words, together with a formula for writing it.

A Conference Abstract

Background information: The process of submitting a paper to a conference

1. The general details of a conference are first made known by means of brochures and notes in journals.
2. If you (or your supervisor) register your interest, you will receive a *Call for Abstracts*. The abstract is used to judge whether you will be invited to attend the conference.
3. If you are invited to attend the conference, you will then later receive the *Call for Papers*.
4. You will be told whether your presentation is oral or a poster.
5. Occasionally, abstracts that have been rejected for the conference (i.e. you won't be asked to attend) are still made available to the conference participants, so that they know what work is being done elsewhere. These are sometimes called *Unpublished Abstracts*.
6. The collected papers – usually from all of the participants, but sometimes only from selected ones – will be published in the conference proceedings either as a hard copy or in a digital format.

Formatting and appearance of the conference abstract and paper. Conference organisers strive for a uniform appearance; the *Instructions to Authors* will give precise requirements for font type and size, margin size and so on.

Be sure that you meticulously follow the *Instructions to Authors*. Even if you think you can improve the appearance by using a different font or size from those that are stipulated, don't do it. You don't want your work to stand out in any way other than in its content.

Purpose of a Conference Abstract

1. **Initial purpose: To enable the conference organisers to decide whether to invite you to present your work at the conference.** There are several aspects to this. They need to decide the following:
   a. Whether your work fits in with the theme of the conference
   b. Whether your work is good enough
   c. If you are invited to the conference, whether you will be asked to present your work orally or as a poster presentation

2. **Purpose during the conference: To enable conference participants to decide whether your work is of interest to them**. They may then do the following:
   a. Attend your oral presentation *or* seek out your poster.
   b. Try to make personal contact with you during the conference.
   c. Contact you after the conference, if they haven't met you during it.

How to write a Conference Abstract (two to three pages)

To get the organisers' invitation to attend the conference, you have to present the **problem, your methods and your results** clearly and in enough detail, *without any other supporting information.*

This is hard discipline: it's not easy. You have to write a self-contained, mini-paper. Therefore, the following are important:

- You need to strip away the detail and decide on the hard-core material of your work.
- Avoid writing in descriptive terms (see Descriptive or Indicative *Abstract/Summary*, page 55).
- You will not be able to fit in more than one or two small figures or tables.
- You need to restrict the number of references, so that the *List of References* does not take away too much of your limited space.
- Think of it as a short story with a clear logical flow: a beginning (*the context and motivation*), a middle (*methods and results*) and an end (*the conclusions or outcome*). Each one of these parts needs to be clearly defined and signalled and in the correct order.
- Use the same formula for the structure and signalling words as for a journal paper *Abstract* (see *Abstract,* Chapter 6: *A Journal Paper, page* 83).

Possible headings

Short abstract (100 words to about half a page): no headings.

Longer conference abstract (two to three pages): Conference organisers usually require the standard *TAIMRAD* (*Title, Abstract, Introduction, Methods, Results and Discussion*) structure of the journal paper.

Section	Cross-Reference to Detail in This Book
Title	See **Title**, Chapter 2: *The Core Chapter*, page 19
Authorship and Affiliation	See **Authorship and Affiliation**, Chapter 6: *A Journal Paper*, page 87
Abstract *(very brief)*	This chapter
Keywords (*possibly*)	See **Keywords**, Chapter 5: *A Journal Paper*, page 87
Introduction	See **Introduction**, Chapter 2: *The Core Chapter*, page 28
Materials and Methods	See **Materials and Methods**, Chapter 2: *The Core Chapter*, page 36
Results	See **Results**, Chapter 2: *The Core Chapter*, page 37
Discussion (*or* combined *Results and Discussion* section)	See **Discussion**, Chapter 2: *The Core Chapter*, page 38
List of References	See **List of References**, Chapter 15: *Referencing*, page 167.

Common mistakes in Abstracts or Summaries

- Too long and too detailed
- Conversely – in the attempt to cut it down to the required number of words – large editorial inconsistencies (usually gaps in the logical flow of the information)
- Vague, imprecise information
- No clear statement of the main problem
- No clear description of the methods
- Non-quantitative description of the results
- A descriptive abstract that describes only the structure of the document and gives no real information

For common mistakes in a journal paper, see *Common Mistakes, Abstract*, Chapter 6, *A Journal Paper*, page 90.

An Executive Summary

Purpose

- To provide a document in miniature that may be read instead of the longer document. The *Executive Summary* is directed at managerial readers who may not read the whole report and who may not have the appropriate technical knowledge.
- To explain your work in terms understandable by the non-expert reader.

Length

An *Executive Summary* is longer than the conventional abstract/summary (apart from some conference abstracts); typically it is 10–25% of the whole document.

Format

- **Unlike a standard (short) summary, it can be organised under appropriate headings and numbered blocks of information and be highlighted in boldface.**
- It should be formatted for accessibility of information and the speed and convenience of the reader.

Structure and Content

- The structure should follow that of the body of your report.
- Although the body of your report may contain technical or scientific terminology, the *Executive Summary* should, as far as possible, be written in non-expert terms.

Checklist for Summary/Abstract/Executive Summary

- ☐ Does it include the following (if appropriate to the subject matter):
  - ☐ A statement that places your work in context
  - ☐ Your method of investigating it
  - ☐ Your main results or observations
  - ☐ Your main conclusion(s)
  - ☐ Your main recommendations(s)
- ☐ If these are inappropriate to your subject matter, does it give *real* information?
- ☐ Have you avoided writing a descriptive type of abstract/summary?
- ☐ Is there any information that doesn't appear elsewhere in the document? If so, incorporate it somewhere.
- ☐ No tables, figures, literature references? (*Except for a Conference Abstract, see below.*)
- ☐ Was the abstract/summary written as the final stage of the document?
- ☐ ***A Conference Abstract***
  - ☐ Does it conform to the conference guidelines (page number and so on)?
  - ☐ Have you included only one or two small illustrations?
  - ☐ Is there a clear logical flow: a beginning (*the context and motivation*), a middle (*methods and results*) and an end (*the conclusions or outcome*)?
  - ☐ Have you included an appropriate number of citations and a short *List of References*?
- ☐ ***An Executive Summary*:**
  - ☐ Does it give the information appropriate to a managerial readership?
  - ☐ Is it written so that a non-technical person can understand it?
  - ☐ Does it follow the structure of the main document?
  - ☐ Does it have appropriate descriptive headings and numbered blocks of information, and is it highlighted in boldface?
  - ☐ Is it written for accessibility of information, and the speed and convenience of the reader?
- ☐ ***A journal paper Abstract*:** See Checklist, *Abstract*, Chapter 6: *A Journal Paper*, page 90.

4 A Literature Review

This chapter covers:

- When you are likely to have to write a literature review
- Purpose of a literature review
- Common difficulties
- What makes a good review?
- General guidelines
  - Step 1: Consulting a librarian about searching techniques
  - Step 2: Using bibliographic data management software
  - Step 3: Searching for tertiary sources (books)
  - Step 4: Searching for secondary sources (review articles)
  - Step 5: Searching for primary sources (e.g. journal papers): choosing your KEY papers
  - Step 6: Selecting the information from KEY papers
  - Step 7: Choosing the tentative topic headings for your review
  - Step 8: Sorting your information into the various topic headings
  - Step 9: Selecting the FRINGE papers
  - Step 10: Selecting, analysing and filing information from the FRINGE papers
  - Step 11: Re-reading your original review papers
  - Step 12: Writing up the literature review
    Connect all material for each topic together
    Connect all topics under series of headings
    Keep re-working it
    Make sure the review is not just who-did-what-and-when lists
- A possible structure for a literature review
- Why your initial review won't be good enough for a thesis
- Common mistakes
- Checklist

What is a Literature Review?

A literature review is an account of what has been published on a topic by established scientists. The publications should be in accredited sources such as journal and conference papers. The review should convey the knowledge and ideas that have been established on the topic and also what their strengths and weaknesses are.

Writing for Science and Engineering.
DOI: http://dx.doi.org/10.1016/B978-0-08-098285-4.00004-2

When you are Likely to Have to Write a Literature Review

1. As an assignment, probably before you start your research work for your thesis or project. Preparing a literature review is often one of the first things a supervisor asks a new graduate student to do. It can seem overwhelming, especially when you've had an initial look at the volume of literature or perhaps the lack of it.
2. As part of a longer document such as a thesis or major report.
3. As an initial scoping for possible thesis topics.

Purpose of a Literature Review

To show that you have a good understanding of the background of your topic of research or investigation. To do this, you need to do the following:

1. Give a coherent account of the various areas of research relevant to your topic.
2. Give a historical account of its development. Its history may span many years or very few, if it is a recently developed area.
3. Show that you know who has done the relevant work, by citing at the appropriate points in the text the names of the authors and the years in which the work was published.
4. Show the links between the various areas of the body of knowledge – the correlations, contradictions, ambiguities and gaps in the knowledge.
5. Show the weaknesses of other work and techniques.
6. Provide a summary of available techniques and materials.
7. Show how your work will form an original contribution.

Common Difficulties

1. Feeling overwhelmed by the quantity of literature.
2. Sometimes having to cope with a lack of literature.
3. Getting started. It is often very difficult to know where to look in the literature to get the general overview you need to start.
4. Knowing how broad or how narrow to make your review.
5. Knowing which documents to discard and which to keep. Your understanding increases over time, and it's important not to discard works that initially don't seem relevant. As understanding develops, documents need to be constantly reviewed.
6. Having the mental discipline to constantly re-read documents to gain fresh understanding.

What Makes a Good Literature Review?

A good literature review presents the facts, but in addition it should also go behind the facts. It needs to:

- Show the issues that have been dealt with in the past.
- Show the issues that are currently being addressed and those that need to be.

- Show the correlations, contradictions, ambiguities and knowledge gaps that exist.
- Show the conflicts between competing research groups.
- Give an analysis and commentary that makes it clear that you understand the issue.
- Show that you are imposing your view on the issue.

A poor review is just an account of who did what and when it was done, without comment on relevance and quality. By doing this, you don't show your competence and involvement; you show that you haven't fully understood the real purpose of a literature review. It happens when by making the following errors:

- You believe that it's not up to you to comment.
- You believe that your role is that of a neutral observer.
- You don't understand the topic sufficiently.

General Guidelines

1. **Do not worry about the sheer volume of literature** that you know exists or that your supervisor may have given you a few specialised papers to start you off.
2. **Be systematic**. It is all too easy to jump around when doing a search. You need to be focused and to keep good records.
3. **You are going to have to be cleverly selective about your choice of literature to search for.** At some early stage, you begin to realise that the process can grow exponentially and become unwieldy. Each review paper that you read may have a multitude of references cited in it. Each journal paper will have anything from a dozen to 50 or more. You can't possibly follow up each one.
4. **It is essential to view it as an iterative process.** During the searching, extracting and filing of material, your understanding of the topic will increase. You won't be able to complete it in only one pass.
5. **Steps 1–12** shown schematically in Figure 4.1 give a series of steps designed to enable you to understand the process of searching for material and consolidating it into a literature review.

   The scheme is a simplified version of the real process, which is obviously more complex and iterative.

Step 1: Consult a librarian about searching techniques

If you don't know how to search efficiently, ask for a librarian's advice, and also attend one of the courses in searching techniques that university libraries offer graduate students.

To search efficiently, you will need to know the following:

- **The relevant searching strategies for your topic.** There are many library databases. You'll probably need more than one of them to do an adequate search.
- **How to do an efficient keyword search of the library databases.** A librarian will be able to help you define your subject in terms of keywords. Inefficient choice of keywords and combinations of them can result in a lot of wasted time and effort.

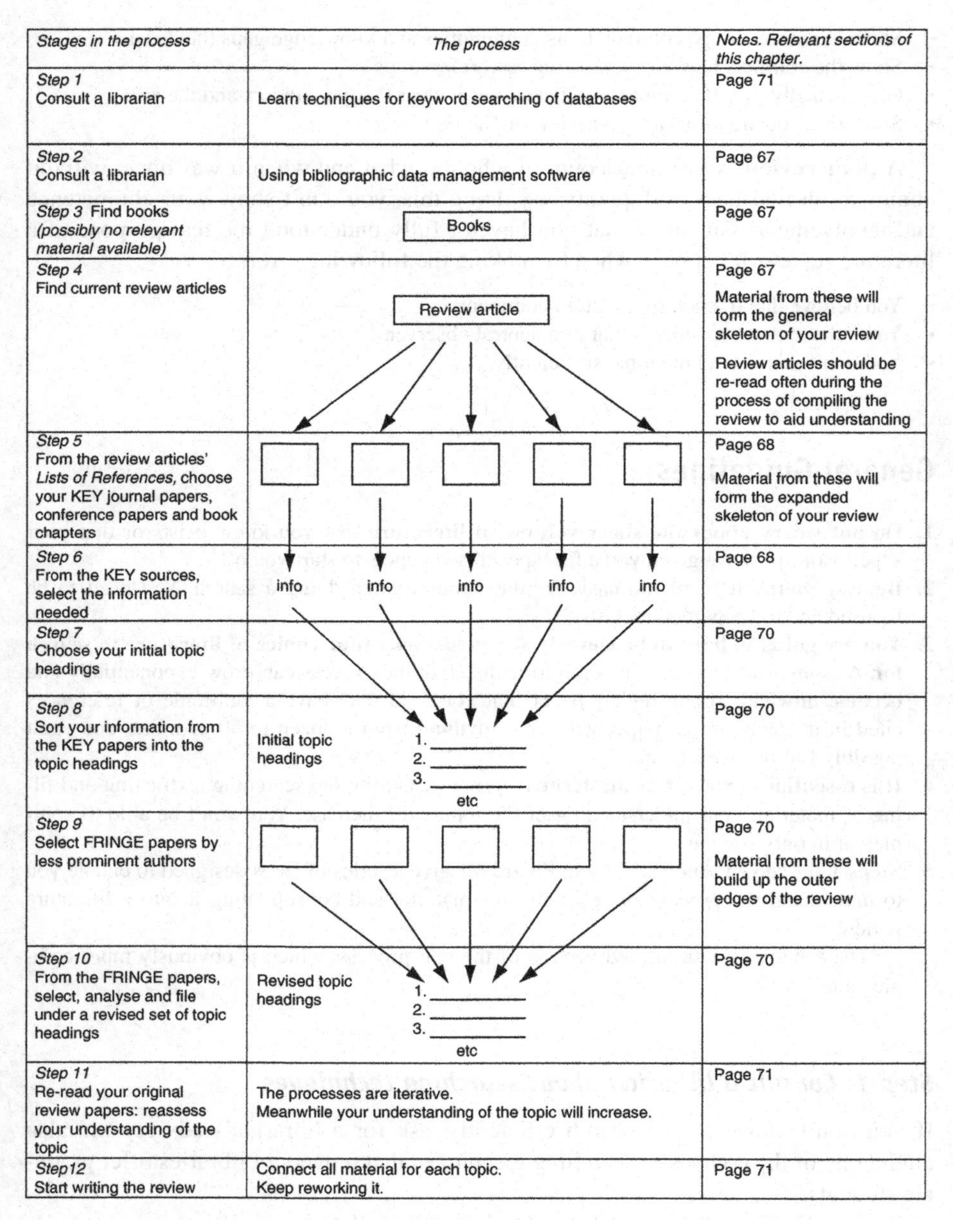

Stages in the process
The process
Notes. Relevant sections of this chapter.
Step 1
Consult a librarian
Learn techniques for keyword searching of databases
Page 71
Step 2
Consult a librarian
Using bibliographic data management software
Page 67
Step 3 Find books (possibly no relevant material available)
Books
Page 67
Step 4
Find current review articles
Review article
Page 67
Material from these will form the general skeleton of your review
Review articles should be re-read often during the process of compiling the review to aid understanding
Step 5
From the review articles' Lists of References, choose your KEY journal papers, conference papers and book chapters
Page 68
Material from these will form the expanded skeleton of your review
Step 6
From the KEY sources, select the information needed
info
info
info
info
info
Page 68
Step 7
Choose your initial topic headings
Page 70
Step 8
Sort your information from the KEY papers into the topic headings
Initial topic headings
1.
2.
3.
etc
Page 70
Step 9
Select FRINGE papers by less prominent authors
Page 70
Material from these will build up the outer edges of the review
Step 10
From the FRINGE papers, select, analyse and file under a revised set of topic headings
Revised topic headings
1.
2.
3.
etc
Page 70
Step 11
Re-read your original review papers: reassess your understanding of the topic
The processes are iterative.
Meanwhile your understanding of the topic will increase.
Page 71
Step12
Start writing the review
Connect all material for each topic.
Keep reworking it.
Page 71

Figure 4.1 Schematic of the stages in writing a literature review. The scheme as shown is a simplified version of the real process, which is obviously more complex and iterative.

Step 2: Using bibliographic data management software

Before you start searching for sources, work out how you will keep records of them. Establish what types of bibliographic data management software are available, e.g. EndNote or RefWorks. Free software is also available, such as Zotero or Mendeley.

Invest the time to learn how to use such software to effectively search, record and create a customised *List of References.* The time saved and the organisational capabilities it will give you are well worth the small effort at the beginning of your searching.

Searching for the primary, secondary and tertiary sources. Using the library databases, there are three broad types of sources you need to search for: tertiary, secondary and primary sources.

1. Tertiary sources are **textbooks.**
2. Secondary sources are **review articles.**
3. Primary sources are the original accounts of the investigations, particularly **journal and conference papers.**

Guidelines for searching for these three types of sources are given in Steps 3, 4 and 5 below.

Step 3: Searching for tertiary sources (books)

At the beginning of your search process, look for **tertiary sources**, i.e. textbooks. You won't find the depth of detail that you will eventually need, but they can sometimes be useful in giving an overview of a field. Some textbooks become acknowledged as authorities and can be acceptable as references.

Many libraries have a large stock of e-books; make sure you search these too.

Step 4: Searching for secondary sources (review articles)

Now search for **secondary sources**, i.e. review articles that are as current as possible, written by accredited top scientists in your particular field. Review articles are summaries of information gathered from primary sources.

The *Bibliography* and citations given in your tertiary source books may lead you to the review articles you'll need. If not, ask a librarian how to search for them.

From a good review article, you should be able to establish the following:

- The general overall sense of the topic.
- The areas that are relevant to your topic.
- Who are the most active researchers, and who are the authorities. Those frequently cited are the most prominent.
- The papers that are regarded as the most important and fundamental, i.e. the most frequently cited sources.

Material from the review articles will give you a general skeleton to build on.

Step 5: Searching for primary sources (journal papers, conference papers, book chapters): choosing your KEY material

Now search for the **primary sources.** These are the first-hand accounts of investigation: journal papers, conference papers, book chapters, theses and reports.

From the large number of citations in the *Lists of References* in the review articles, you can now select the KEY papers:

- The most frequently cited **papers** in the review articles.
- The most frequently cited **authors** in the review articles.
- The **titles** that are the most relevant to your topic.
- The **most recently dated**. Later in the process, to show the historical development of the topic, you may need to access papers that are older.

Material from the KEY papers will form the expanded skeleton of your literature review.

Concentrate at this stage on identifying the KEY papers. If during this part of the search process, you find a paper that doesn't seem to be directly relevant, file it in another place. You may later find that it can become one of your FRINGE papers (see Step 9, *Select the FRINGE Papers*).

Step 6: Select the information you need from the KEY papers

Choose sources that will enable you to answer the following questions:

- How does this topic fit into a broader view of the research area?
- Why is it an important topic?
- What is known about the topic?
- What is ambiguous, in dispute, unknown? Why?

If you are writing a literature review as anything other than a preliminary overview of an area of work, you will also need to answer the following:

- Why do these gaps in the knowledge need to be filled?
- Which gaps do you propose to fill in your research?
- Why have you chosen them?
- How do you propose to do it?

Why These Questions Need to be Answered

It is important that these questions are answered as well as you can because it is very easy to forget how much you know and how little your readers know in comparison. When we forget this, we tend to think that the facts will speak for themselves – that if they are just presented, they will form a coherent account. But they rarely do.

A good analysis of the literature should set the whole context of your work and demonstrate your competence in the area.

1. First read the abstract.
   a. **Prioritise your papers initially by reading the abstracts.** From the abstract, you can establish the paper's relative relevance to your needs.
   b. **Then establish from the abstract which parts of the paper you will need to read.** For instance, if you need details only of the method, it could be a waste of time to read the whole article in detail.
2. Read the relevant part of the paper.
   Then skim-read the rest of the article. From this, you can establish whether there is additional relevant information.
3. Interact with a document: select and annotate.
   Even if you select good material while reading a document, you will probably find that when you come to write it up, you will need to look at the original document again. Most people's memories are far from perfect, and explanatory notes that you made at the time of reading will often not make sense when you come back to them later on. There is always something that you will have overlooked, distorted or oversimplified. Therefore, you will save a lot of time later if you highlight or annotate by any method you like.
4. Classify your key findings under major headings.
   As you gather and analyse your sources, think of the categories into which you can classify them. You can find that each fits into a specific category – or more than one – and that there is some logical organisation taking shape in your mind. As you analyse even more papers, you'll find that your major categories can be broken down into subcategories. Eventually you should be able to pull them all into a coherent whole (see Step 7, *Choose the Topic Headings*).
   The following types of things should be noted:
   - Key elements of the original data and wording
   - An overall summary of the document
   - Key discussion points from the original document
   - Keywords
   - Your own comments and queries

   Make enough notes to jog your memory later.
5. If you find an article difficult to understand:
   Put it aside for a while and come back to it. There are two reasons why you may be finding it difficult:
   1. Your understanding of the topic is not yet sufficient. Later in the process you may be able to understand it with no difficulty.
   2. It may not be your fault that it is tricky to understand; remember that a hallmark of a good writer is the ability to explain even complex topics clearly. If the text seems muddy, it could be that the article is not well written.
      Distinguishing between whether your understanding is lacking or whether the article is poorly written at this early stage can be difficult. Keep the paper and come back to it later.
6. Read all of the important documents at least twice; come back to each one at a later stage in your work.
   You will find extra material and insights in a paper when you re-read it sometime later. At this time, you will have a better grasp of your area of research and of the place of your own work within it. The following are useful things to do at this time:
   1. Note the significant points.
   2. You are probably also now in a better position to check calculations. You may be surprised by the number of errors it's possible to find. You will add depth to your review by discussing them.

3. You may now have the experience to be able to ask such things as the following:
- Why didn't they do a particular experiment?
- Is that conclusion sufficiently well supported?
- Are the statistics weak?

When you have developed more of the depth of knowledge necessary, you may find that the data in some of the published material are somewhat inconclusive. Discussing this in an informed and objective way will add depth to your review.

Step 7: Choose the tentative topic headings for your review

1. Work out the topic headings.

- Do not think that these headings should be fixed. As the review evolves, you will probably find that new topics emerge, and others become less significant or merge with others.
- Make your topics specific. Instead of having a small number of overall, unspecific topics such as 'Issues associated with bacterial adhesion', think of a number of subtopics, each one explicitly stating the specific issue. *For example:*

  Methods for investigating
  Historical background
  Standard techniques
  Current technology

2. Establish either one folder containing these topic headings *or* separate folders for each topic.

If your review is going to be a major work, separate folders are usually more convenient. You will file information under the headings after reading and extracting information from each source.

Under these headings, also note cross-references to various other papers. This saves having to copy large amounts of information; at the final stage of sorting information, you will come back to these papers and extract from them the information needed.

Step 8: Sort your information into the various topic headings

- Your information will come from the following:
  - The material you have placed on file (either in one large folder or in separate ones under various topic headings)
  - The papers' abstracts
  - Your own comments
- Sort your information into the relevant heading or file.
- If some papers span more than one topic, duplicate the material to each heading or into each file. You can decide later where it best fits.

Step 9: Select the FRINGE papers

- These are the papers by the less prominent authors.
- Find them from:
  - The work cited in your KEY papers.

  - Library database searching. From your initial reading, you should now be able to compile the keywords you need for efficient library database searching. If you have problems, consult a librarian.
- Assess the research philosophies, the scientific rigour of the techniques, the results and the interpretations.

Material from the FRINGE papers will build up the outer edges of your review.

Step 10: From the FRINGE papers, analyse and file the information under headings as you did for the KEY papers

Keep revising your categories for both KEY and FRINGE papers as your understanding of the topic grows. Consolidate them as necessary.

Step 11: Re-Read your original review papers: reassess your understanding of the topic

- You will find that your perspectives will change and your understanding will increase if you approach this as an iterative process.
- Repeat the re-reading and analysis of your review papers and your other important papers during the whole process of assembling the material for your review.

Step 12: Write up the literature review as a final stage

Don't try to write up the literature review before you have assembled all of the information. The more you investigate, the more you will understand about the topic; a review written up too early is likely to need much rewriting.

The steps to take from here are the following:

1. Connect all the material for each topic together:
   a. Take the material under each heading. Sort and re-sort it.
   b. Look for possible subheadings
   c. Look for similarities, contrasts, inconsistencies, gaps in the knowledge, links between the topics and subtopics.
   d. Write the text to link these ideas together.
2. Connect all of the topics together under a series of headings.
   Work out a logical order in which to place your separate topics.
3. Keep reworking it.
   It may take many reworkings to produce a coherent review. The material within each topic and the overall structure of the document will need a lot of work before it is satisfactory.
4. Make sure that the review is not made up of who-did-what-and-when lists, with no comment on relevance and quality.
   For example: Brown (20xx) showed that… and Smith (20xx) found that…
   A good review should provide an analysis and commentary that makes it clear that you understand the issue.

5. Compile the *List of References*
The cross-referencing between your text and the *List of References*, together with the details needed for listing each reference, is riddled with convention. See Chapter 15: *Referencing*, for details.
A data management software package such as EndNote will do this cross-referencing automatically and is well worth using.

A Possible Structure for a Self-Standing Literature Review (i.e. not part of a thesis)

Section	Comments	Cross Reference to Detailed Material
Title	In contrast to the title of most other documents that you may have to write, the title of a review will be a general description of the research area	See *Title*, Chapter 2: *The Core Chapter*, page 19
Abstract	In contrast to most other Summaries/Abstracts that you may have to write, a short *Summary* in a review may need to be a description of the document structure rather than give only informative material. But give informative conclusions if possible	See **The different types of content in an abstract/ summary**, Chapter 3: *An Abstract, a Summary, an Executive Summary, page 55*
Introduction	Provide the following information: • The general historical development of the topic • The various areas of the topic • The document structure in the final paragraphs	See **Introduction**, Chapter 2: *The Core Chapter*, page 28.
Sections appropriate to the subject matter		
Conclusions	Summarise the various conclusions, including any contradictions, ambiguities or gaps in the knowledge	See **Conclusions**, Chapter 2: *The Core Chapter*, page 39
List of References		See Chapter 15: *Referencing*, page 169

Why Your Initial Literature Review Won't Be Good Enough for a Thesis

If you are writing a thesis or another major body of work, your initial literature review – done at the start of your study – can never be good enough to be the version incorporated into your final document.

The understanding that comes through time, the re-reading of the material and the discovery of new material will result in a final literature review that will be very different from your first. Make sure that the review contained in your thesis is written up as a final stage of putting your thesis together.

Supplementary Tabulated Presentation for a Thesis

Consider supplementing the literature review material with a tabulated presentation to summarise the content of each of the relevant papers.

See Chapter 13: *Thesis*, page 143.

Common Mistakes

1. Finishing the literature review before you have fully understood the issue.
2. Just giving an account of who did what and when. An inadequate literature review is one that is little more than a who did what and when list.
3. No obvious logical thread running through the whole review and through each category in it. This can be the result of point 2.
4. Not pointing out the gaps in the knowledge and any ambiguities. Not becoming involved.
5. Referencing errors (see *Common Mistakes,* Chapter 15: *Referencing*, page 188).

Checklist for a Literature Review

- ☐ Does your review show the issues that have been dealt with in the past?
- ☐ Does it show the issues that are being and need to be currently addressed?
- ☐ Does it cite the key reviews on the subject? The KEY papers? The more FRINGE papers?
- ☐ Does it show the correlations, contradictions, ambiguities and gaps in the knowledge?
- ☐ Does it show the conflicts between competing research groups?
- ☐ Does it give an analysis and commentary that makes it clear that you understand the issues?
- ☐ Does it avoid giving just an account of who did what and when?

5 A Research Proposal

This chapter covers:

- Types of proposals:
  - As an assignment at the start of graduate work
  - To a funding body
  - To a commercial organisation
- Writing for the non-expert reader
- Writing for the commercial sector
- Checklists

This chapter is written as a short basic guide. There is a wealth of online information, particularly for grants aimed at European Union (EU) organisations, and US organisations such as National Institutes of Health (NIH), National Science Foundation (NSF), Defense Advanced Research Projects Agency (DARPA) and others.

Types and Purpose of Proposals

1. **As a possible assignment at the start of postgraduate work.**
   Purpose: So that your supervisor can see that you have a clear idea of previous work in the area, the research problem and the procedures you will use to tackle it.
2. **Together with the supervisor, as a proposal to a funding body or an outside organisation to persuade them to fund your research.**
   Purpose: To convince the funding or commercial organisation that your work will be of value to them and to persuade them to fund it. In some cases, this can occur after you have already been working on the research topic for some time.

How to Write It

As an assignment at the start of your postgraduate work

You need to clearly explain the following:

1. The objectives of your proposed research
2. Previous work in the area
3. How you are proposing to tackle it

Writing for Science and Engineering.
DOI: http://dx.doi.org/10.1016/B978-0-08-098285-4.00005-4

4. The time frame for each stage
5. Facilities, resources, laboratory equipment and technical help needed

A possible structure

Summary	See Chapter 3: *An Abstract, a Summary, an Executive Summary,* page 53
Research Objectives	See **Objectives**, Chapter 2: *The Core Chapter,* page 30
Literature Survey or ***Background*** Supervisors will realise that at this early stage, you will not have come to grips with very much of the literature. But they will expect a clear explanation of the general framework of the research that has been done in your area, together with appropriate specific work.	See Chapter 4: *A Literature Review,* page 63 and **Background**, Chapter 2: *The Core Chapter,* page 30
Materials and Methods or ***Procedures*** This will need a description of the expected stages of the research and an outline of the techniques you expect to use during each one. It may be effective to describe each expected stage and its procedures under an appropriate series of headings. **Tense of the Verb** Use the **future** form. Note: A *Materials and Methods* or *Procedures* section in other types of papers uses the past tense: **The bacteria <u>were cultured</u> on both solid and liquid media (*past tense*).** In a research proposal: **The bacteria <u>will be cultured</u> on both solid and liquid media (*future tense*).** **For tightly defined topics** (such as those for some master's level theses or smaller projects, e.g. diploma projects): You and your supervisor may already know almost exactly how you are going to tackle the project. It will be relatively straightforward to explain this. **For those topics that are less well defined** (such as Ph.D. projects and projects where you will follow research leads and possibly construct equipment or devise methods of which you may not have any clear idea at present): State clearly how you propose to tackle the first stages of the project. Then follow with a reasoned description of the framework that the research is likely to follow and the possible procedures that may be needed. *Example:* **The initial stage of this study will be made up of…This will be followed by…If it is found that…, the next stage will consist of…**	See **Methods**, Chapter 2: *The Core Chapter,* page 36 and **Methods**, Chapter 6: *A Journal Paper,* page 83

If needed: ***Schedule of Tasks*** or ***Time Management*** or ***Expected Time Frame***	See *Schedule of Tasks/ Time Management*, Chapter 2: *The Core Chapter*, page 33
Resources The facilities, resources (including interactions with other organisations), laboratory equipment and technical help needed.	See *Requirements*, Chapter 2: *The Core Chapter*, page 35

A Proposal to a Funding Body or to a Commercial Organisation

General criteria by which an application for financial support is judged:

1. The validity of the central concept.
2. The soundness of the experimental design.
3. The significance of the research.
4. The relevance to the funding organisation's programme.
5. Your competence and that of the other personnel who will be involved.
6. The adequacy of the research facilities.
7. The appropriateness of the budget. (Remember that too modest a budget proposal can be as damaging as an overblown one; it shows your poor judgement.)
8. If appropriate: the validity of the evaluation mechanism. The more novel the project, the more it will need an effective programme for evaluating it.

A proposal to a funding organisation

Application forms are usually provided; therefore, the recommended sequence of sections is automatically determined.

Otherwise, the standard *TAIMRAD* structure of a journal paper is acceptable (*Title, Abstract, Introduction, Materials and Methods, Expected Results, and Discussion*), with emphasis on the **significance of the proposed work with respect to the concerns of the funding body**.

In addition to the above:

- Evidence of your ability to carry out the work (your position, publications, honours and awards)
- Facilities available to you (including interactions with other organisations)
- Cost estimates

There are a few fundamental guidelines to remember:

1. **Follow the instructions immaculately.**
   Read them thoroughly a number of times, and keep referring to them while you write the application.
2. **Don't exceed the stipulated total length.**
   Don't be tempted to think that exceeding the word limit is justified in your case because the proposed work is so important or too complex to fit into it. Moreover, don't consider

reducing the font size or line spacing to fit more in. You need to compete on the same basis as other applicants, and deviations from the rules will only irritate the assessors.

3. **Design your application with both specialists and non-specialists on the committee in mind.**
   Embedding your detail within a framework of cleverly designed headings, subheadings and listed points will make it much more easily accessible to all your assessors, both specialist and non-specialist. It is a much greater achievement to be able to design a readily navigable document with a clear logical pathway – the red thread – through it, than to bombard your assessors with solid detail.
4. **Don't skimp on research design and methods: get expert help on study design.**
   This material will be assessed very critically by your specialist assessors. Make sure that you describe not only your methods but also the design of your experimental work, i.e. why you choose to tackle it this way.

Assessment by the funding organisation of your proposal

Most funding organisations will probably judge you on the following seven points:

1. **What you want to do**: a listed series of objectives.
2. **Why you want to do this research**: the context of the work and the gap in the knowledge that you're aiming to fill.
3. **What you have done already**: your preliminary results described so that they are relevant to your stated objectives.
4. **How you are going to do it**: you research design and methodology.
5. **Possible conclusions** to be drawn from your proposed work.
6. **What your results will be useful for**: the possible applications of your proposed work in relation to the funding body's interests.
7. **Your track record, of you as a person or as a team.**

Assessment criteria: what you need to cover

1. **Rationale for research**
   Why should the research be done?
   What will it address?
   How does it fit into the research landscape?
   Include statement of purpose/research aims, hypothesis, new knowledge, Technical advance, innovation
2. **Research design and methods**
   Scientific protocol, sample recruitment and characteristics
   Feasibility
   Study methodology
   Validity of the data (make sure results would be significant)
   Timelines
3. **Anticipated outcomes/impact on the goals**
   How will the research contribute to one of the goals?
   How will it address the problem or contribute new knowledge?
   How will knowledge be transferred to end users?
4. **Track record of the research team**
   Describe relevant track record to show the team can deliver the proposed outcomes
   Highlight key skills for expertise
   Justify staff roles

A research proposal to a commercial organisation

There are two aspects that will influence the way you write your proposal:

1. The main concerns of the organisation's personnel will be their business plan and whether your work will contribute to the company's competitiveness and profitability. They may have little or no interest in the academic implications of your proposed work.

Some large companies are wealthy enough to be able to fund "blue-skies" research, knowing that they will eventually be able to use the intellectual information to contribute to their wealth. However, many companies have to concentrate on their immediate or mid-term plan; they will therefore be more receptive to projects that will require minimum additional development and/or commercialisation costs and will provide quick returns to the company.

2. Your report will need to be understood by people with no expertise in your particular field.

There may be no one who is familiar with the basic knowledge and terminology of your subject. Even in companies that have the expertise, your report may be passed on to people such as financial personnel.

Your report should therefore:

- Sell your research to the organisation, without misrepresenting, exaggerating or appearing pushy.
- Emphasise the potential advantages of your research to the organisation's profitability.
- Assure the organisation that they will have full and uninterrupted access to your progress at all times.
- Use a *Glossary of Terms* to clearly explain terminology that may not be familiar to the organisation.
- Be written in language that does not need expert knowledge to be understood. But it shouldn't be oversimplified.
- Be very concise, well presented and clearly worded, without elaborate justifications or full descriptions of complicated techniques.
- Be easy to navigate through, with a clearly defined pathway that enables a non-expert to understand the material. See **The Importance of Overview Information: Building a Navigational Route through Your Document**, Chapter 1: *Structuring a Document*: *Using the Headings Skeleton*, page 11.

Questions to ask yourself

While you are putting together the proposal, you need to ask yourself:

- What is the commercial significance of my proposed research?
- What questions will they expect to be answered?
- How can I write this so that it will be understood by a person without specialist knowledge in my field?

Depending on the type of organisation, you may need to avoid the classic, scientific *TAIMRAD* format of *Title, Abstract, Introduction, Materials and Methods, Results and Discussion,* and aim instead for a structure that is more suited to a commercial organisation.

Suggested structure for a proposal to a commercial organisation

You may not need all of the following sections.

Cover Page Includes the name of the organisation to which the proposal is being submitted and title of proposal: **Proposal to** (*description of what you propose to do*)	See **Title Page**, Chapter 2: *The Core Chapter*, page 21
Executive Summary (on a separate page following the cover page) A summary written in non-specialist language outlining what you propose to do and how it would benefit the company.	See **Executive Summary**, Chapter 3: *An Abstract, a Summary, an Executive Summary*, page 53
Research Objectives Clearly and briefly describe the aim of the research. In presenting the focus from the company's point of view, do not try to anticipate your results. Just say what you are aiming to do At the end of this section state that: • **Results will be presented in the form of a report to be used by (*name of the organisation*).** • **The research personnel will be available for discussion on any part of the document for a mutually agreed upon period of time after completion of the report.** • **Submission of the final report will be approximately (*x months*) after the start of the project.** **Present the expected stages of the work in sequence** At the end of each stage, use a subheading ***Outcomes***; in this section, say what you expect the outcome of each stage of the work to be. This does not mean anticipating the results; it means stating that at this point, you will be able to say, for example, which one of the several growth media is the most efficient at promoting cell growth.	See **Objectives**, Chapter 2: *The Core Chapter*, page 30
Schedule of Tasks or ***Time Management*** or ***Expected Time Frame*** State your expected time schedule of the following: **1.** The various tasks (possibly with a Gantt chart, *see Figure xx*, page xx) **2.** The reports that you will write for the organisation, e.g.: • Preliminary report: 3 months • Interim report: 6 months • Final report: 9 months	See **Schedule of Tasks/ Time Management**, Chapter 2: *The Core Chapter*, page 33

Expected contents of the final report **1. State what you expect the final report to contain.** ***For example:*** • **The report will present the results of (...), together with analysis and discussion appropriate for consultancy purposes.** • **The report will discuss (*the various aspects of the experimental work*).** • **The report will discuss the design, operating and maintenance strategies of (*equipment that you may be developing*).** You may also need to state that: **2. The report will be finalised in consultation with members of (*name of the organisation*). This ensures that the organisation knows that they will be involved in the final version of the report.** **3. Discussion of the results and oral presentation of the work will be available on request. This ensures that the organisation knows that they have full and uninterrupted access to your progress at all times.**	
Requirements A description of what you expect to need, other than money, from your funding organisation during your research.	See **Requirements,** Chapter 2: *The Core Chapter*, page 35
Costs A description of the money you expect to need from your funding organisation during your research. This should be discussed and agreed on with your supervisor/immediate superior/university commercialisation division before submission of the proposal to the potential funding organisation.	See **Costs,** Chapter 2: *The Core Chapter*, page 36
Ownership/Confidentiality An agreement between you and the commercial organisation funding you that gives you some right of publication of your results, while assuring the organisation that you will not divulge commercially sensitive information.	See **Ownership/ Confidentiality,** Chapter 2: *The Core Chapter*, page 35
Evidence of your ability to carry out the work Your position, publications, honours, awards.	

Checklist for a Research Proposal at the Start of Academic Work

Does the proposal contain the following:

☐ A clear explanation of the general framework of the previous research in your area, together with appropriate specific work

☐ A clear statement of your objectives

- ☐ The expected stages of the research and the expected methodology for each one
- ☐ A description of the time frame for each stage
- ☐ Facilities, resources, laboratory equipment and technical help needed

Checklist for a Proposal to a Funding Body or a Commercial Organisation

- ☐ Does the title give immediate access to the main point of the proposal?
- ☐ Does the *Executive Summary* or *Abstract* give an accurate and informative overview of the whole document, without being vague?
- ☐ Are your research objectives clear?
- ☐ Does the *Introduction* briefly identify the following:
  - ☐ The critical problems *and*
  - ☐ Your main purpose
- ☐ Do you show all of the facilities and resources needed for the project?
- ☐ Do you describe the expected stages of the research and the expected outcome of each?
- ☐ Do you describe your qualifications and those of other personnel for carrying out this work?
- ☐ Do you show the expected time frame for completion of the following:
  - ☐ The next stage
  - ☐ The whole project (if necessary)
- ☐ Do you itemise the costs as accurately as possible?
  - ☐ Have you underbudgeted?
  - ☐ Is your budget overblown?
- ☐ **For a funding organisation:**
  - ☐ Do you show the relevance to its overall programme?
  - ☐ Have you explained the rationale for research?
  - ☐ Have you adequately described your research design and methods?
  - ☐ Have you given the anticipated outcomes/impact on the goals?
  - ☐ Have you given details of the track record of the research team?
- ☐ **For a commercial organisation:**
  - ☐ Do you emphasise the potential advantages to the organisation's commercial activities?
  - ☐ Do you show that you are going to fill a need for the organisation?
  - ☐ Does it assure them that they will have full and uninterrupted access to your progress at all times?
  - ☐ Is it written in terms that can be understood by a person without specialist knowledge in your field?
  - ☐ Do you clearly show what is expected from the organisation?
  - ☐ Are ownership and confidentiality issues adequately addressed?

6 A Journal Paper

This chapter covers:

- The general structure of a journal paper
- How to start writing a journal paper
- How to write the following sections:
  - Title, Abstract, Running Title, Authorship and Affiliation, Keywords, Abstract, Introduction, Materials and Methods *or* Procedure, Results, Discussion, Results and Discussion, Conclusions/Conclusion
- The process of publishing a paper
- General guidelines for figures
- Collected checklists
- Planning a journal paper: Question sheet

The General Structure of a Journal Paper

A journal paper will usually follow the classic *TAIMRAD* basic skeleton of sections (*Title, Abstract, Introduction, Materials and Methods, Results* and *Discussion*) in its general format. Many papers will need these actual sections: some journals with high impact factors, e.g. *Nature* and *Science*, will require a more narrative structure but will need to follow the basic skeleton's scheme in its overall plan.

The following elements are generally found in a journal paper:

Title (and **Running Title** if needed)
Authors(s) and Affiliation(s)
Abstract
Keywords
Introduction
Materials and Methods
Results
Discussion
Sometimes: **Results and Discussion** combined as one section
Sometimes: **Conclusions, Recommendations**
Acknowledgements
List of References

Writing for Science and Engineering.
DOI: http://dx.doi.org/10.1016/B978-0-08-098285-4.00006-6

The following table shows the purpose of each element. See also Chapter 2: *The Core Chapter*, for further information that is less specific to a journal paper.

Section	***Purpose in a Journal Paper***
Title	To adequately describe the contents of your document in the fewest possible words.
Running Title (if needed)	The short title required by journals for the tops of the pages. Running titles can use abbreviations.
Authors(s) and Affiliation(s)	To show the people who did the work presented in the paper, the institutions where it was done and, if necessary, the present addresses of the authors.
Abstract	To give the reader a *brief* overview of all of the key information in the paper: objective, methods, results, conclusions.
Keywords	This is a short list of words relevant to your work that will be used by electronic services.
Introduction	• To clearly state the purpose of the study. • To allow readers to understand the background to and motivation of the study, without needing to consult the literature themselves. • To indicate the authors who have worked or are working in this area, and to describe their chief contributions. • To indicate correlations, contradictions and gaps in the knowledge, and to outline the approach you will take with respect to them. • To provide a context for the later discussion of the results.
Materials and Methods	To describe your experimental procedures. Aim: repeatability by another competent worker.
Results	To present your results but not to discuss them.
Discussion	To show the relationships among the observed facts that you have presented in your paper and to draw conclusions.
Results and Discussion	This section is a useful way of structuring the results and their significance.
Sometimes: **Conclusions**	To give an overview of the conclusions that you have already drawn previously in the paper.
Recommendations	To propose a series of recommendations for action.
Acknowledgements	To thank the people who have given you help in your work and in the preparation of your paper.
List of References	A list of the works that you have cited in the text. Strong conventions govern this process.
Illustrations (Figures and tables)	

How to Start Writing a Journal Paper

Most supervisors will recommend the following sequence of stages:

1. **Choose the journal.**
2. **Establish the particular type of document you are writing, and the journal's allowed number of figures, tables and words/characters for that type.**

3. **Work out the illustrations – the figures and tables –** that you want to include, particularly those in the *Results* section. Many science graduate students are happier to think more in images than words, and this first step enables you to start planning the logical thread in terms other than the written word. Each figure should convey at least one obvious take-home message of your paper.
4. **Then write a 'working abstract' with a clear logical thread running through it.** This won't be the final abstract for your paper: write it without a word limit but keep it brief. This will force you into creating the logical pathway of your results and the conclusions you want to bring out from them. This is the main part of your paper, the part that your readers will be most interested in. The logic needs to be clearly thought through before you start to write anything else.
5. **Create the final versions of your illustrations, figure titles and captions (legends).** Make sure that the important points are all clearly contained in the illustrations. It's well known that readers of a paper look at the figures very early in the process of reading a paper. Carefully consider which information is to be placed in the main body of the text and which in the figure captions (legends). Consider using schematics to give an overview of a complex hypothesis or procedure.

   Test Points 4 and 5 before you go any further: the entire logical story of your paper should be clear from your working abstract and your illustrations.
6. **Write the *Methods* section.** Most people find this the easiest section to write and always start the writing process here. Supervisors may say that they want you to write this section while you are doing the experiments; however, most graduate students are reluctant to do this, preferring to write only when they must. Moreover, if it's written too early, you may find it needs to be rewritten. It is always much more difficult to do a major rewrite than to write it when you are more sure about what is needed.

See also: *Planning a Journal Paper: Question Sheet*, this chapter, page 108.

Guidelines: Writing the Various Elements of a Journal Paper

This section gives suggestions – some of them are almost formulae – for writing the various sections or elements of a journal paper. These formulae are not necessarily followed in all good papers, but they are designed to give you an objective way of obtaining an effective structure.

Also listed are the common mistakes that graduate students most dislike in published papers that they need to read. You'll already realise that in some papers, the information is more difficult to access than in others. This is often due to poor structure or lack of specific information in that section.

We'll deal with these sections or elements in the usual order in which they appear in a paper.

Journal Paper Title

Purpose of the title

- To give readers immediate access to the subject matter.
- To give informative, not generalised, information: to reflect the important content of the paper.
- To describe the contents of the paper in the fewest possible words. Note: This can lead to clumsy strings of nouns (see *Noun Trains*, Chapter 18: *Problems of Style*, page 214).

Formulae for effective titles

Ask yourself these two questions:

What is the single most important point made in this paper?
How would I convey that to another scientist in one short sentence or phrase?

Stating the conclusions in the title

Many experimentally based papers, particularly those in the biological and medical sciences, use a short, declarative sentence that gives the major conclusion. Some authors like them, some don't; take advice from your supervisor. This type of title has a verb in it (underlined in the following examples):

Serotonin neuron transplants exacerbate l-DOPA-induced dyskinesias in a rat model of Parkinson's disease
TAK1 inhibition promotes apoptosis in KRAS-dependent colon cancers

However, effective titles can be very informative without giving the major conclusion:

A multi-axial fatigue model for fibre-reinforced composite laminates based on Puck's criterion
Regulation of circadian behavioural output via a microRNA-JAK/STAT circuit

Using a colon in a title: a hanging title

Readers' preferences for colons in titles vary. However, if you have lot of information that you need to include in a title, you can try making it more readily understandable by using a colon. In such a title (a 'hanging title'), one part of it – usually the first – describes the general area of work, the other part gives the specific information.

Inelastic interface damage modelling with friction effects: application to Z-pinning reinforcement in carbon fibre epoxy matrix laminates
First part is the general area; the second part is the specific information.

Trying to include all of this information without the colon makes it much more difficult to read:

The application to Z-pinning reinforcement in carbon fibre epoxy matrix laminates of inelastic interface damage modelling with friction effects

Ineffective titles often too generalised

They don't give immediate access to the main point. Here is a series of titles describing the same information, ascending from too general (not useful) to highly specific (very effective):

Genetic control of changes in root architecture
Too generalised, not enough information.

Genetic control of nutrient-induced changes in root architecture
Improved slightly, with one additional item of specific information.

An Arabidopsis MADS gene that controls nutrient-induced changes in root architecture
Fully informative, with very specific information.

Title: common mistakes

- Does not give immediate access to the subject matter of the paper.
- Too generalised: not enough information for readers to assess whether they need to read it.
- Too much detail.
- Too long and clumsy; made up of long strings of words (noun trains, see page 214) that are awkward to read or even ambiguous.

Title: checklist

- ☐ Does it give the reader immediate access to the subject matter?
- ☐ Is it informative, not generalised: Does it reflect the important content of the paper?
- ☐ Is it too long/too detailed?
- ☐ Is it too short and uninformative?
- ☐ Is the wording clumsy? Does it make sense?
- ☐ Is its meaning absolutely clear?

Running title

This is the short version of the title requested by some journals. It will appear at the top of the page of the inner pages of the printed paper. The journal will limit the number of characters. In the running title, you can use abbreviations that you may not want – or be allowed to use – in the main title:

***Main title*: Facile atom transfer radical polymerisation of methoxy-capped oligo (ethylene glycol) methacrylate in aqueous media at ambient temperature**
***Running title*: ATRP of methoxy-capped OEGMA**

Authorship and Affiliation

Purpose

To show the people who did the work; the institution(s) where it was done; current addresses of authors.

For formatting, see the journal's *Instructions to Authors.*

Keywords

This is a short list of words relevant to your work, required by only some journals. These will be used by readers to search online.

It is important to work out the most likely keywords a potential reader might use to search for information. They should include both general and specific items.

***Title*: Subretinal electronic chips allow blind patients to read letters and combine them into words**
***Keywords*: Subretinal neural-prosthetics; retinal implant; retinitis pigmentosa; blindness; artificial vision; bionic vision**

Abstract

For more general material about Abstracts, see *Abstract*, Chapter 2: *The Core Chapter*, page 54.

Journal paper Abstract: purpose

1. Students usually say that an abstract's main function is to save time, determine whether it is relevant to their interests or help them decide whether they need to read the whole paper. That's true, but an often unrealised function of a well-written abstract is that it helps the brain assess and better understand the complex material in the rest of the paper. Efficient readers will always read the abstract first.

2. Effective abstracts have become crucial in a journal paper. People obtain the main points of your work from your online abstract, and from that information, they decide whether they need to read the whole of your paper. If the *Abstract* is weak, your work won't get the advertising that it may deserve.

An abstract should therefore be a miniaturised, fully informative version of the whole paper.

Journal paper Abstract: ineffective abstracts

Abstracts that are not fully informative are not suitable for a journal paper abstract. See Chapter 3: *An Abstract, a Summary, An Executive Summary*, page 54, for a description of the differences between the informative abstracts needed in journal papers and the uninformative, descriptive abstracts used in review papers.

Journal paper Abstract: difficulties in writing

- Deciding on the core information
- Making sure that all aspects are covered
- Linking the information up into a coherent story
- Making the final reduction to the required word count

Journal paper Abstract: format

The most usual format is from 200 to 300 words as one paragraph of continuous text. Some journals require their abstracts to have a given set of headings, e.g. *New England Journal of Medicine*, which requires Background, Methods, Results and Conclusions.

Journal paper Abstract: how to write it

General guidelines

1. Make the description of the methods brief unless you are presenting a new method. Many excellent abstracts do not mention well-established methods.
2. The results usually make up most of the *Abstract.* Make them as quantitative as possible.
3. Do not use non-standard abbreviations, use them only if they are widely recognised in your field.

Journal paper Abstract: formula for an effective structure

Many excellent abstracts do not slavishly follow this formula. However, if you are relatively new to the process of writing a journal paper, this formula will give you a straightforward structure for an effective *Abstract*:

1. **A brief statement of the context** (the *specific field* of your work)
2. **The gap in the knowledge that your work aims to fill** (*why* you are doing the work, the *motivation*)
3. **A brief description of your methods** (often not needed) (*how* you did the work)
4. **Your results**, which should take up most of the *Abstract* (*what* you found)
5. **Clearly signalled main conclusion** (the *overall significance* of your work)

Then ask yourself: Is it a miniaturised, fully informative version of the whole paper?

Example

We'll take a very clearly structured brief abstract. It is only 124 words but clearly shows all five elements listed above, apart from the methods. It uses signalling words or phrases for the gap in the knowledge (*why* you are doing the work, the *motivation*), the results and the main conclusion.

A reader will find that an abstract structured and signalled in this way is very intelligible and useful.

The generation of protective memory-like CD8+ T cells during homeostatic proliferation requires CD4+ T cells

Antigen-specific memory T cells are a critical component of protective immunity because of their increased frequency and enhanced reactivity after restimulation.	First sentence: brief statement of context.
However, it is unclear whether 'memory-like' T cells generated during lymphopenia-induced homeostatic proliferation can also offer protection against pathogens.	Second sentence: signals the gap in the knowledge. **Signalling phrase: *However, it is unclear***
Here we show that homeostatic proliferation-induced memory (HP-memory) CD8(+) T cells controlled bacterial infection as effectively as 'true' memory CD8(+) T cells, but their protective capacity required the presence of CD4(+) T cells during homeostatic proliferation. The necessity for CD4 help was overcome, however, if the HP-memory CD8(+) T cells lacked expression of TRAIL (tumour necrosis factor-related apoptosis-inducing ligand, also called Apo-2L).	Third and Fourth sentences: the results. **Signalling phrase: *Here we show***
Thus, like conventional CD8(+) memory T cells, the protective function of HP-memory CD8(+) T cells shows dependence on CD4(+) T cell help.	Final sentence: clear statement of main conclusion. **Signalling word: *Thus***
124 words	Note: No methods

Hamilton, S. E., Wolkers, M. C., Schoenberger, S. P., Jameson, S. C. (2006). The generation of protective memory-like CD8+ T cells during homeostatic proliferation requires CD4+ T cells. Nature Immunology 7, 475.

Here is a list of the signalling words and phrases useful in a journal paper *Abstract*. They should appear in this order at the beginning of the sentence or near to it:

1. **Brief statement of context**	No signalling words
2. **Gap in the knowledge** (*why* you are doing the work, the *motivation*)	*However...* *It is unclear...* *It is not yet known...* or combinations of them
3. **Results** (*what* you found out)	*Here we show...* *The results show that...* *It was found that...*
4. **Main conclusion** (the work's *significance*)	*Thus...* *Therefore...* *It is concluded that...* *We conclude that...* *Our results suggest...*

Journal paper Abstract: common mistakes

- No clearly stated conclusions.
- No clearly obvious purpose or motivation of the study.
- The information is too generalised: vague and imprecise.
- No obvious structure; an illogically presented story.
- Non-quantitative description of the results.
- Too many small facts, unnecessary details.
- Unfamiliar abbreviations.
- Important information often missing, e.g. a clear description of the methods, a quantitative description of the results.
- A descriptive abstract (see Chapter 3: *An Abstract, a Summary, an Executive Summary*, page 55): describes only the structure of the document, and gives no real information.

Abstract: checklist

- ☐ Is it a miniaturised, fully informative version of the whole paper?
- ☐ Is it really informative? No use of descriptive phrases?
- ☐ Is the main conclusion clearly stated?
- ☐ Is the story presented logically?
- ☐ Is there a clear logical flow: a beginning (*the context and motivation*), a middle (*methods and results*) and an end (*the conclusions or outcome*)?
- ☐ Does it contain the sort of information that a reader doing an online search would like to find?
- ☐ Have you avoided using non-standard abbreviations?
- ☐ Are there no citations?
- ☐ Is it too short and uninformative?

Journal Paper *Introduction*

Introduction: general guidelines: how to write it

1. **How long should it be?** Without a coherent plan, many people spend far too long writing an *Introduction* section and then find that it needs to be severely cut. Remember, the *Introduction* section in a journal paper is not intended to show the breadth and depth of your knowledge, as expected in a thesis.
2. Clearly show the **main objective** of the work. In a journal paper, this is best done in the last or next to last paragraph of the *Introduction* (see below).
3. **Review the literature and show the relationships between the various areas of work.** Show the contributions of others, with reference citations of their work. The references that you cite should be carefully chosen to provide the background information relevant to your paper.
4. **Show where there are correlations, contradictions, ambiguities and gaps in the knowledge.**
5. **Define the specialist terms used in the document.** The *Introduction* is the correct place for definitions of terms. Don't assume that if you've defined them in the *Abstract*, you don't need to define them again in the *Introduction*.
6. **Structure the *Introduction*.** The *Introduction* tells a story. It should have an obvious logical flow running through the development of the information in it (see box, *Introduction: Formula for an Effective Structure*, below).

Journal paper Introduction: formula for an effective structure

If you are relatively new to the process of writing a journal paper, this formula will give you a straightforward structure for an *Introduction*. However, experienced authors in cutting-edge science papers may not strictly follow points 1 and 2 of this formula but instead may choose to blend them.

1. ***The Beginning*:**
   Briefly summarise the relevant current knowledge: link it together as a coherent story and support it with references as necessary.
2. ***The Middle*:**
   Now move on to the gap in the knowledge: areas where there is less or no knowledge, or where the evidence is conflicting. This should follow logically from the material in Point 1 (The Beginning). The same signalling words can be used for the gap in the knowledge as in the *Abstract*:
   However…
   It is unclear…
   It is not yet known…
   or combinations of them
3. ***The End*:**
   (a) **State the objective of your work.** In the final paragraph or next to last paragraph, make a brief, clear statement of the objective of your work, i.e. the gap in the knowledge that your work is meant to fill and the research question specific to the work in the paper. Make the objective arise out of the gap in the knowledge; don't state it clumsily, e.g. *The reason for doing this study was…*

(b) **Then, briefly summarise your approach.** Many excellent papers do this. This is an effective way of rounding off the *Introduction* and moreover helps the reader's understanding of the work.

(c) **If appropriate, briefly summarise your results.** Some people approve of making a brief statement of the results; some do not. Check it with your supervisor.

Signalling phrases for the results (as in an *Abstract*):

Here we show…

The results show that…

It was found that…

Journal paper Introduction: how to start writing it

Using the formula above, here are guidelines for writing the *Introduction* so that the logical thread is clear:

1. Start by writing the last paragraph first (The End, Point 3 above): the research question you are answering followed by your means of answering it.
2. Then in this last paragraph, identify all of the concepts that you'll need to develop in the preceding paragraphs (Points 1 and 2 above). Thus, your final paragraph will determine the logical thread that will need to be developed in the first part of the *Introduction*.

Journal Paper Introduction: Tense of the Verb

See *The Correct Form of the Verb*, page 224, Chapter 18, *Problems of Style*, for guidelines on using tense in technical documents, together with examples of the various forms of the tenses of the verb. See also Appendix 2: Tenses and forms of the verb, page 261.

You should use a mixture of present and past tenses in their various forms, because you are describing both the established body of knowledge (*present tense*) and what people have discovered (*past tense*).

If you have doubts, read it aloud to yourself and use your instinct about whether to put it in the past or the present tense, and the appropriate form of the verb to use. It's often surprisingly accurate.

Example of past and present tenses in a journal paper *Introduction*

Restoration of function… has been explored (*past*) in clinical trials in patients with advanced Parkinson's disease (PD). The results have been (*past*) highly variable, with some patients showing a substantial recovery in motor function, and others showing little or no improvement. These discrepancies have been suggested (*past*)… to be attributable to the differences in dissection and

preparation of the fetal tissue, in which tissue clumps, tissue stripes, or single-cell suspensions have been used (*past*). It is known (*present*) that transplanted ventral mesencephalic tissue… contains (*present, established knowledge*) also other neuronal cell types… and variations in tissue dissection are (*present*) likely to result in varying numbers of different cell types in the graft cell preparation. The serotonin neurons are (*present*) of particular interest in this regard because they have the capacity to…

Adapted from Carlsson *et al.*, Serotonin Neuron Transplants Exacerbate L-DOPA Induced Dyskinesias in a Rat Model of Parkinson's Disease. *The Journal of Neuroscience*, 2007, **27(30):** 8011–8022; doi: 10.1523/JNEUROSCI.2079-07.2007.

Journal paper Introduction: common mistakes

- No clearly stated aim/objective/purpose of the study.
- The literature has not been adequately reviewed. For example: the pivotal references may not have been cited; only a few references may have been cited for a thoroughly researched area of work; the gaps or inconsistencies in the knowledge may not have been clearly pointed out and so on.
- Too detailed, rambling, unspecific, unstructured, irrelevant material.
- Too long.
- Too short and general.
- Same as other papers: the material and the references cited are frequently identical to those in papers from the same work group.
- If the author/date citation system is being used (see Chapter 15: *Referencing*, page 169), sentences are split by large numbers of references.

Introduction: checklist

- ☐ In the final paragraphs, does it *clearly* state the objective of the study?
- ☐ Following the objective, does it give a summary of your approach? Possibly: the results?
- ☐ Does it adequately review other people's work?
- ☐ Does it identify the correlations, contradictions, ambiguities and gaps in the knowledge in this area of research?
- ☐ Does it place your study into the context of other people's work?
- ☐ Does it have a clear logical structure with an obvious red thread of logic running through it: a beginning, a middle and an end?
- ☐ Does it define the specialist terms?
- ☐ *If appropriate*, does it give a historical account of the area's development?

Methods (also called Materials and Methods or Procedure)

Purpose

- To describe your experimental procedures
- To give enough detail for a competent worker to repeat your work

- To describe your experimental design
- To enable readers to judge the validity of your results in the context of the methods you used

Difficulties

- **Not many.** Describing experimental methods is usually very straightforward.
- **Therefore, it is often the best place to start writing.** Writing a document is often difficult, and it's not usually the best tactic to write it in sequence from beginning to end. Start with the section that is going to give you the fewest problems. **See *How to Start Writing a Journal Paper*, page 84.**

How to write it

- **You should have already worked out the illustrations needed in your paper** (see *How to Start Writing a Journal Paper*, page 84). Make sure that the description of your methods is consistent with any of the illustrations in this section, e.g. schematics, tables.
- **Logically describe the series of experimental steps so that the whole procedure could be repeated by a competent worker in your field.** You need to tread a fine line between giving the right amount of detail for a colleague and giving the sort of superfluous detail that such a person would not need. Think in terms of describing only the essentials.
- **Ask yourself whether you might be too familiar with the techniques.** You might make the mistake of leaving out elements of procedure descriptions that are essential but which you take for granted. If you think this is the case, give your description to a colleague to read.
- **You also need to give the rationale behind your experimental design**. This section should not just be a list of the experimental steps you took. A reader must be able to understand from the *Introduction* and the *Materials and Methods* sections why you chose to do it this way.
- **Make sure that you don't introduce some of the results.** It is quite easy to do this accidentally. The *Methods* and the *Results* sections need to be very strongly separated from each other in their contents.
- **Should it be written in chronological order?** Inexperienced people sometimes present this section chronologically. However, sometimes it is only convenience that determines the order of experiments. The methods section should not be a diary of what you did; it should have a logical flow.
- **How much detail?**
  Established techniques don't need a detailed description; however, novel techniques or variations of a previous one do. Don't forget to state sources of chemicals, model numbers of equipment and so on; these are important in evaluating the results, e.g.

  The surgery was performed under injectable anaesthesia (20:1 mixture of fentanyl and Dormitor, intraperitoneal; Apoteksbolaget, Stockholm, Sweden).

- ***We* is usually acceptable. *I* is rarely acceptable.** The occasional use of *We* in an active construction is acceptable, e.g.

  Using detailed 3D kinematics and body mass distributions, we examined net aerodynamic forces and body orientations in slowly flying pigeons (*Columba livia*) executing level 90° turns.

Beware, however, of using *We* too often. It could sound child-like: *We did this, then we did that.*

- **Referencing in the *Materials and Methods* section.** If you have to refer to literature to explain a technique, give enough brief information for the reader to get an outline of the technique.

Good*: *Cells were broken by ultrasonic treatment as previously described (Smith, 20xx).
Poor*: *Cells were broken as previously described (Smith, 20xx).

- **If you cite a previous paper for a method, make sure that it does indeed contain a description of that method.** Many postgraduates complain of citations that refer back to a previous one for details of the technique, only to find that this citation doesn't give the details either – and so on backwards through the literature until it all fades away.

Tense of the verb

See *The Correct Form of the Verb*, page 224, Chapter 18, *Problems of Style*, for guidelines on using tense in technical documents, together with examples of the various forms of the tenses of the verb. See also Appendix 2: *Tenses and forms of the verb*, page 261.

For experimental work: Use a form of the past tense. You are describing work that you did.

Correct: **Nickel ammonium sulphate (2.5 mg/ml) was used (*past tense*) to intensify the staining.**
Incorrect: **Nickel ammonium sulphate (2.5 mg/ml) is used (*present tense*) to intensify the staining.**

For description of morphological, geographical or geological features: Use the present tense.

The eucervical sclerites are connected to the postcervical sclerites, each of which is differentiated (*present tense*) into a relatively hard sclerotised base and a flexible distal part. All three paleosols show (*present*) a greater degree of development than the surface soils. Better development is displayed (*present tense*) in terms of greater clay accumulation, higher structural grade, harder consistency and thicker profiles.

Materials and Methods/Procedure: common mistakes

- Not enough critical detail to enable someone unfamiliar with the method to repeat it. Conversely, too much trivial detail.
- Detailed text, where an illustration would be more appropriate.
- Illogical description. This can happen when several procedures are described together.
- Being referred back to the literature with not enough summarised information to be able to recognise the technique. *For example:*... as described previously (Brown, 20xx).
- Citing a paper for a technique that does not contain a description of that technique.
- Introducing some of the results.

Checklist for the Materials and Methods/Procedure

- ☐ Does it provide enough information to allow another competent worker in your field to repeat your work?
- ☐ Is all of the necessary detail given about the equipment used, e.g. the model number of an instrument?
- ☐ Are there no detailed descriptions of standard instrumentation and techniques?
- ☐ Are the necessary details provided for the following:
  - ☐ Modifications to standard instrumentation and techniques
  - ☐ New techniques
  - ☐ Any organisms used, e.g. species, variety, age, weight
- ☐ Does it state precise treatment/drug regimens?

Results

Purpose

- To present your results but not to discuss them
- To give readers enough data to draw their own conclusions about the meaning of your work

Difficulties

- Deciding how much detail to include

General comments

- **Your results need to be clearly and simply stated.** This is the core section of the paper: the new knowledge that you are presenting to the world.
- **It needs to be presented as a logical story.** If it is interrupted by material that is too detailed or is not directly relevant, your readers are going to become disorientated and lose the thread.
- **The need for excellent illustrations.** Readers familiar with the topic will usually visit the illustrations in the *Results* section very early in the reading process. This highlights the need for illustrations to be as self-explanatory as possible, with informative titles and captions (legends).

How to write it

- **Write your *Results* section around the illustrations. They will have already been planned in detail** (see *How to Start Writing a Journal* Paper, page 84, and *Designing Figures and Tables*, Chapter 2: *The Core Chapter*, page 47). You should have already ensured the following:
  - They have been carefully chosen to illustrate the points you are trying to make.
  - The titles and captions (legends) make the illustrations as self-explanatory as possible.
  - Each one conveys at least one obvious take-home message.
  - They have been prepared according to the journal's instructions.
- **In the text, describe the most important aspects of the results.** You need to guide the concerning reader what to look for in the tables and figures. A *Results* section is not just made up of a series of graphs and tables; there must be explanatory text linking the illustrations.

- **Amount of detail:** It should not be a detailed diary-like account of your results. In any piece of research, there will inevitably be results that are not worth presenting.
- **Dealing with repetitive data.** Do not give them all. Present representative data, and state that they are representative. If needed, present the other data in the journal's online supplementary information.
- **It is important to include anomalous results that do not support your hypothesis.**
- **Do not discuss the results.**
- **The *Results* section is the next easiest section to write, after the *Methods* section.** It is efficient to be able to write the *Results* when you have finished writing the *Methods*. See *How to Start Writing a Journal Paper*, page 84.

Results: common mistakes

- Illustrations that are not self-explanatory.
- Inadequate titles and captions.
- Poorly selected illustrations that do not illustrate the main results.
- Illustrations not prepared according to the journal's instructions.
- Inadequate explanation in the text. The main points are not well described; readers are left to deduce the results from only the illustrations.
- Repetition in the text of large amounts of material that are already shown in the figures and tables. Only key material should be pointed out in the text.
- Too much detail. Readers do not need every item of data collected.

Tense of the verb

See *The Correct Form of the Verb*, page 224, Chapter 18, *Problems of Style*, for guidelines on using tense in technical documents, together with examples of the various forms of the tenses of the verb. See also Appendix 2: *Tenses and forms of the verb*, page 261.

Use the past tense. You are describing the results you obtained.

Example:

In the DA-wide grafts, there was (past tense) an enrichment of serotonin neurons compared with the DA-narrow group.

Checklist for the *Results*

- ☐ Are your illustrations well chosen, i.e. to show your most important results?
- ☐ Are the illustrations presented well and self-explanatory as far as possible, with thoughtfully written titles and captions (legends)?
- ☐ Is there explanatory text pointing out the key results and trends?
- ☐ Have you avoided giving a diary-like account of the data?
- ☐ Have you avoided discussing the results?

☐ If you have repetitive data, have you included only representative data in the *Results?*
☐ Are the figures prepared exactly according to the journal's *Instructions to Authors*?

Journal Paper *Discussion*

Purpose

- To state clear conclusions
- To show the significance of your results and how your results lead to your conclusions
- To give the answer to the gap in the knowledge – the research objective – stated in the *Introduction*
- To explain how your results support the answer
- To show the relationships among your observations
- To put the results into context, particularly with reference to other people's work

Difficulties

- Being unsure where to start
- Now knowing what or how much to put in it
- Challenge of getting a logical flow

Overall guidelines: how to write it

1. In a good *Discussion*, you present the *significance* of the work described in the rest of the document. You don't just restate the material; you need to answer the question:
   How does the work described in this paper add to the existing body of knowledge?
2. Your conclusions need to be clearly stated.
3. Most authors believe that conclusions are best placed towards the end of the *Discussion*; you therefore present your material and build up to them. However, other authors start the discussion with their conclusions and then present the evidence for them. If you are not yet experienced in writing a paper, you are likely to find it more acceptable to place them at the end and build up to them.
4. Show how your results and interpretations agree or contrast with previously published work.
5. Each conclusion should have a sound basis of evidence. Make sure that each one is well supported by the facts.
6. Describe other studies' findings accurately, fairly and objectively.
7. Don't be afraid to defend your conclusions. But in doing this, treat other studies with respect.
8. State any limitations of your methods or study design.
9. State any important implications of the work.
10. Make sure there is a clear logical thread through it. It should not be an unstructured brain dump.
11. Point out in your own work any exceptions or any lack of correlation, and define unsettled points.

    Be open and honest about inconsistencies or gaps in the data. Never try to cover up data that do not quite fit; it will be obvious to an expert reader that you are fudging. Do not avoid mentioning them, either.
12. Avoid any far-fetched hypotheses: keep all your speculation within reasonable bounds.

Journal paper Discussion: formula for structure

Lead readers through a logical sequence as follows:

1. Provide a very brief statement of the objective of the work, i.e. the research question you stated in the final paragraphs of the *Introduction*.
2. Include a very brief statement of your results. Some authors prefer the results not to be restated here; check with your supervisor.
3. Provide a very brief statement of the conclusions that don't need discussion, followed by those that do. *Numbers 1, 2 and 3 should* ***together be very succinct, i.e. no more than one paragraph.***
4. Discuss the significance of your results and put them into context of other people's. Show how your results and interpretations agree or contrast with previously published work.
5. Build up in the final paragraph(s) to a summary of all the conclusions, very clearly stated.
6. State any implications of your work. You may also like to include the direction of your future work. However, check this with your supervisor; because of scientific competitiveness some do not like this to be given.

Word choice in a journal paper Discussion

- ***Prove*** **is too strong a word**; apart from mathematical proofs, nothing can be assumed to be fully proven in science. Your assessors will prefer you to state your conclusions less equivocally. Instead, use **show, demonstrate** or ***indicate.*** To a native English speaker, *indicate* has slightly less positive ring than *show* or *demonstrate*, but to all intents, the three words are synonymous.
- ***Appear*** sounds weak but can be used so that it is equivalent to *show/demonstrate/indicate* as in the following:

  Thus, CD9 appears to be essential for sperm–egg fusion.

  This is a positive statement and is more acceptable to a journal than the unequivocal:

  Thus, CD9 is essential for sperm–egg fusion.
- ***Reveal*** has an element of showmanship display to it and has been overused by the tabloid press. It is therefore best to avoid frequent use.
- **Hedging words: *May be, might be, could be, probably, possibly*** It is acceptable to use hedging words; science is rarely cut-and-dried.

  But don't go to extremes of hedging:

  Acceptable: **These results suggest that A is the cause of B.**

  Acceptable: **These results suggest that A may be the cause of B.**

  Too cautious: **These results suggest the possibility that A may be the cause of B.**

Discussion: tense of the verb

See *The Correct Form of the Verb*, page 224, Chapter 18, *Problems of Style*, for guidelines on using tense in technical documents, together with examples of the various forms of the tenses of the verb. See also Appendix 2: *Tenses and forms of the verb*, page 261.

As in the *Introduction*, the verbs here should be a mixture of past and present tenses. If you have doubts, read it aloud to yourself and use your instinct. It's often surprisingly accurate.

Use the past tense for results (yours and those of others), but use present tense for established fact, to describe existing situations and for your answers to the research question.

> *Example*
> **The bacterial biofilms were found (*past: your results*) to vary in structure over time. A possible reason for this variation is (*present*) that they could have been (*past*) subject to predation. The presence of protozoa has (*present: established fact*) a significant impact on biofilm structure. For example, Brown (20xx) found (*past*) that the numbers of protozoa increased (*past: other people's results*) in mature biofilm…**
>
> **The results of this study suggest (*present*) that *Nitrosomonas* species are (*present: your answer to the research question*) slow-growing and very sensitive to environmental change.**

Discussion: common mistakes

1. **Main conclusions not clear.** A poorly planned *Discussion* runs the risk of obscuring the main conclusions of the work.
2. **Conclusions are insufficiently supported.** Unspecified assumptions are made, leading to unjustified conclusions. Each conclusion that is drawn should have a sound basis of evidence.
3. **The significance of the material is not adequately discussed.** You should not just restate the material; you need to discuss its significance, interpret it in the context of other work and then draw conclusions from it.
4. **Too long, unstructured and wordy.** A *Discussion* should not be an unstructured brain-dump. You need to thoroughly plan the points you want to make and the logical thread running through it, and then state the points as clearly and concisely as possible.
5. **Too short.** If your *Discussion* is described as being too short and limited, this probably means one or both of the following:
   - You haven't thought out all of the implications of your work.
   - You aren't familiar enough with the literature to be able to place your work in context.
6. **You have ignored some of your results or anomalies.** This will be obvious to an expert reader.
7. **Don't hypothesise too wildly.** Make sure that all of your hypothesising is within the realms of possibility.

> **Checklist for the *Discussion***
>
> ☐ Is each conclusion well supported? Do you give sound evidence for each one?
> ☐ Is the whole *Discussion* well structured?
> ☐ Is there a red thread of logic running clearly through it?

- ☐ Does it give an interpretation of the results, rather than a restatement of them?
- ☐ Does it show how your results and interpretations agree or contrast with previously published work?
- ☐ Is it frank in acknowledging anomalies in your work, and are they explained?
- ☐ Is the *Discussion* free of vague statements?
- ☐ Is it accurate, fair and objective regarding other studies' findings?
- ☐ Have you avoided far-fetched hypotheses?

Results and Discussion

A *Results and Discussion* section is often easier to write and more accessible for the reader than separate *Results* and *Discussion* sections.

Formula for an efficient *Results and Discussion* section:

First set of results
  Discussion
Second set of results
  Discussion
  etc.
Overall *Discussion* incorporating the conclusions

Conclusions/Conclusion

Some journals require one or other of these sections; most expect conclusions to be incorporated into the *Discussion*.

Purpose

To present your conclusions and their significance based on the previous material in the document.

How to write it

1. *Important*: there should be no new findings in this section. Each conclusion must be based on material that has already been presented previously in the document.
2. Each conclusion should be related to specific material.
3. Each conclusion should be briefly stated.
4. The *Conclusions* section not only reviews the results or observations but it also interprets them. In this section, you can therefore point out the following:
   - What is important and significant
   - Why the results or observations are valid
   - Any criticisms or qualifications you may have of your own work

5. A numbered or bullet-pointed list can be used if appropriate. Start with your main conclusion, and then present the other conclusions in descending order.
6. A *Conclusion* section is required by some journals. This is usually written as one or two paragraphs (i.e. not as a list) and serves the purpose of rounding off the document and summing up your conclusions and opinions.

Conclusions: *common mistakes*

- **Vague, generalised statements.** Important: you are stating the conclusions you can draw from your results and your interpretation of their significance.
- **Poor or no basis of evidence for each conclusion.** Each conclusion *must* be soundly based on material previously stated in the document.

Checklist for the *Conclusions*

☐ Is each conclusion based on material that appears elsewhere in the document?
☐ Is there a sound basis of evidence for each of your conclusions?
☐ *If necessary*, do you point out the importance, significance, validity, criticisms or qualifications of your work?

Acknowledgements

See *Acknowledgements*, Chapter 2: *The Core Chapter*, page 23.

References

See Chapter 15: *Referencing*, page 169.

Figures for a Journal Paper: General Guidelines

See also *Illustrations*, Chapter 2: *The Core Chapter*, page 44, for further guidelines and comprehensive checklists.

1. **Follow the *Instructions to Authors* very carefully.** Journals give detailed instructions on how figures should be set out.
2. **The journal will not publish the same data in both figure and table form.** Use the most appropriate form.
3. **You may have to submit backup data for figures.** Some journals ask for the data to be submitted in table form as backup to the figures. This gives the reviewers the value for data points, instead of having to interpolate them from the graphs. However, this does not mean

that the data will be published. Check with the *Instructions for Authors* to see if the journal requires this.

4. **Keys.** The *Instructions to Authors* usually include instructions on where to insert the key to a figure. It is usually in the figure itself, but in some journals, it is included in the figure caption.
5. **Use of an already published figure.** See also *Copying and Adapting Illustrations*, Chapter 15: *Referencing*, page 169.

   If you need to reproduce a figure that has already been published by another author, you must get permission in writing to use it. Write to the editor of the journal in which the original publication appeared, asking for permission to reproduce it, and saying in which journal you plan to publish it. Journals usually give such permission. You may be asked to write to the original author as well.

   The journal will usually require you to use, at the end of the caption to the reproduced figure, a standard wording in acknowledgement, e.g. *Reproduced with permission from...*

   Some journals have rules concerning whether figures can be adapted or redrawn in any way. The usual form of acknowledgement is at the end of the figure caption: *Adapted from* (reference), or *Redrawn from* (reference).

The Process of Publishing a Paper

Choosing a journal

We'll assume that your supervisor will have suggested an appropriate journal. If not, you will need to consider the following factors:

- **The journal's impact factor.** This is calculated yearly and is a measure of the frequency that the journal's articles are cited in the literature. It's frequently used as a measure for the relative importance of a journal within its field, with journals with higher impact factors usually judged as more important than those with lower ones.

  The impact factor is based on the number of readers: high impact factor journals publish papers with the broadest general interest and therefore have the greatest number of readers. Journals that are increasingly specialised have a smaller readership and therefore a lower impact factor, but the quality of the papers might well be the same.

  Your supervisor will have the knowledge to choose a journal in which the paper has a reasonable chance of acceptance. If you aim too high, there is a greater possibility of rejection and the repeated revising and resubmission to another journal will increase your frustration and also the waiting time to publication.
- **The level of prestige of the journal.** Those of greater prestige will have a higher standard and a greater rejection rate than those lower on the scale, but acceptance will increase your professional standing.
- **The lead time to publication.** Some journals have extensive delays before publishing. This can be due to a slow refereeing process, editorial delays and the difficulties of production schedules. You can find this out by asking other people who have published in a specific journal or by contacting the editor.
- **Whether the journal has page charges.** Some journals publish papers with no charge to the author. Others have a rate per page (e.g. electronic version of the *Astronomical Journal 2011*, US$110, with pages that use colour at US$350 per page), all of which can mean a substantial cost.

Submitting the manuscript

You'll need to meticulously follow the journal's instructions even if you don't like the results. They will specify things such as the font type and size, margins and equations.

1. ***Style and formatting of the manuscript***
   Journals have reasonably straightforward online instructions to authors. In most journals, they are available online under headings such as Instructions to Authors.
2. **Submitting the manuscript**
   Manuscripts are submitted online. At the time of submission, you will need to send the editor appropriate comments.

The most frequently needed comment is an explanation of the importance of this work. A formula for dealing with the possible other comments is as follows:

1. An introductory short paragraph: Attached is our manuscript (title) by (names). We should be grateful if you would consider it for publication in (name of journal).
2. One sentence each for describing:
   - The current state of the field
   - What you did in your study
   - Your results
   - The significance of your results
3. One paragraph describing why your paper would be of interest to the readers of that particular journal.
4. Suggested reviewers and contact details.
5. If you have competitors, state who you would prefer not to review your work.
6. Concluding paragraph thanking the editor for considering your manuscript for publication.

What happens after submission

This depends on the journal, but in general, the following happens:

1. The editor sends the manuscript to the referees, usually two to four.
2. The referees may take a long time to review it, in some cases, 6 months or more.
3. They will then send their comments back to the editor, together with their recommendations about publication. Each referee may recommend any one of the following:
   - **Acceptance with no revision needed.** This is rare.
   - **Acceptance with revisions** (conditional acceptance).
     The manuscript will be published if the referee's recommended alterations are made. Usually no further experiments are needed. The referees may ask for more information, e.g. about the experiments, or a more detailed *Introduction* or *Discussion*, or a further analysis of data by a specified method.
   - **Rejection with offer to reconsider**, with an instruction that more work is needed.
   - **Rejection because the manuscript is not suitable for publication *in this journal.***

On the basis of all the referees' decisions, the editor then chooses from the following possibilities:

1. Whether to accept your manuscript for publication subject to amendment, *or*
2. To subject it to a further refereeing process after major alteration, *or*
3. To decline it

You will be told the decision in the editor's letter and also be given copies of each of the referees' comments. You will not know the referees' identities. The editor will also send instructions about the resubmission of the manuscript.

How to deal with referees' comments and amend your paper

1. **General advice:** All of the comments should be read with great care. Most referees' evaluations are usually helpful, and you can take advantage of a fresh viewpoint on both your writing and your work. The final version of the paper can be considerably strengthened by using the comments of a good referee.
2. **How to deal with comments that you don't agree with:** If the suggested changes seem unnecessary or, with good reason, unacceptable to you, then the editor can be given a reasoned argument as to why you believe that a particular change need not be made. There are several things you need to address when evaluating a comment from a referee.
   - *Does the fault lie with you or them?* Some referees' comments can show that they have misunderstood or misinterpreted your material. You then have to establish whether this is (1) because you haven't explained it well enough or (2) the referee doesn't know what he or she is talking about. It is easy in the heat of the moment to assume the latter, but it requires careful reflection.
   - *Is the referee possibly not an expert in the field?* The referees may not be the ultimate authorities on your topic. This may be no fault of the editor; it may be difficult to find appropriate referees for each of the hundreds of manuscripts that an editor has to deal with each year.
   - *Is there a political reason for the comments?* Rivalry between research groups and institutions may cause biased refereeing. Your supervisor should be able to advise you on this.
   - *Are the remarks trivial?* In a few cases, you may have just cause to feel peeved. Some referees, if they are unable to make substantial comments, feel the need to justify their appointment by pointing out minor errors such as in the wording. Such comments can often reflect personal quirks and may not make for a valid comment.
   - *Are all of a referee's comments negative?* This may mean that your paper has no worth at all, that the referee is prejudiced or that he or she is trying to impress the editor.

Whatever your conclusions about the referees in terms of these questions, you cannot use words such as *lazy, trivial and useless* in your rebuttal. If you decide not to abide by a referee's suggested amendment, you need to send the editor a calm, well-reasoned and well-written defence that avoids angry terms. Your arguments should be contained in the formal covering comments when you resubmit your amended manuscript to the editor.

The editor will take note of your argument. If your facts are correct and your reasoning is sound, he or she will be able to use your argument as justification for reversing a negative decision.

Proofreading

When your manuscript has been finally accepted, the next stage of the process is to receive the journal's final version to proofread, either as a.pdf or a.doc file, or as online. This needs to be done meticulously, and you may need to use standard proofreading symbols.

If your paper has been rejected

Assess the paper and the rejecting journal's audiences very critically, do more work, alter it so that it is suitable for a different audience and go through the process of submitting it to another carefully chosen journal, probably with a reduced impact factor. Some papers have been submitted a number of times, but obviously you want to minimise this process.

Collected Checklists for a Journal Paper

Title: checklist

- ☐ Does it give the reader immediate access to the subject matter?
- ☐ Is it informative, not generalised: Does it reflect the important content of the paper?
- ☐ Does it describe the contents of paper in the fewest possible words?
- ☐ Is it too long/too detailed?
- ☐ Is it too short and uninformative?
- ☐ Is the wording clumsy? Does is make sense?
- ☐ Is its meaning absolutely clear?

Abstract: checklist

- ☐ Is it a miniaturised, fully informative version of the whole paper?
- ☐ Is it informative (with *real* information), not descriptive?
- ☐ Is the main conclusion clearly stated?
- ☐ Is the story presented logically?
- ☐ Is there a clear logical flow: a beginning (*the context and motivation*), a middle (*methods and results*) and an end (*the conclusions or outcome*)?
- ☐ Does it contain the sort of information that a reader doing an online search would like to find?
- ☐ Have you avoided using non-standard abbreviations?
- ☐ Are there no citations?
- ☐ Conversely, is it too short?

Introduction: checklist

- ☐ In the final paragraphs, does it clearly state the objective of the study?
- ☐ Following the objective, does it give a summary of your approach? Possibly: the results?
- ☐ Does it adequately review other people's work?

- ☐ Does it identify the correlations, contradictions, ambiguities and gaps in the knowledge in this area of research?
- ☐ Does it place your study into the context of other people's work?
- ☐ Does it have a clear logical structure with an obvious red thread of logic running through it: a beginning, a middle and an end?
- ☐ Does it define the specialist terms?
- ☐ If appropriate, does it give a historical account of the area's development?

Materials and Methods/Procedure: checklist

- ☐ Does it provide enough information to allow another competent worker in your field to repeat your work?
- ☐ Is all of the necessary detail given about the equipment used, e.g. the model number of an instrument?
- ☐ Are there no detailed descriptions of standard instrumentation and techniques?
- ☐ Are the necessary details provided for the following:
  - ☐ Modifications to standard instrumentation and techniques
  - ☐ New techniques
  - ☐ Any organisms used, e.g. species, variety, age and weight
- ☐ Does it state precise treatment/drug regimens?

Results: checklist

Do the *Results*:

- ☐ Are your illustrations well chosen, i.e. to show your most important results?
- ☐ Are the illustrations well presented and self-explanatory as far as possible, with well considered titles and captions (legends)?
- ☐ Is there explanatory text pointing out the key results and trends?
- ☐ Have you avoided giving a blow-by-blow account of the data?
- ☐ Have you avoided discussing the results?
- ☐ If you have repetitive data, have you included only representative data in the *Results?*
- ☐ Are the figures prepared exactly according to the journal's *Instructions to Authors*?

Discussion: checklist

- ☐ Is each conclusion well supported? Do you give sound evidence for each one?
- ☐ Is the whole *Discussion* well structured?
- ☐ Is there a red thread of logic running clearly through it?
- ☐ Does it give an interpretation of the results, rather than a restatement of them?

- ☐ Does it show how your results and interpretations agree or contrast with previously published work?
- ☐ Is it frank in acknowledging anomalies in your work, and are they explained?
- ☐ Is the *Discussion* free of vague statements?
- ☐ Is it accurate, fair and objective regarding other studies' findings?
- ☐ Have you avoided far-fetched hypotheses?

Conclusions/Conclusion: checklist

- ☐ Is each conclusion based on material that appears elsewhere in the document?
- ☐ Is there a sound basis of evidence for each of your conclusions?
- ☐ *If necessary*, do you point out the importance, significance, validity, criticisms or qualifications of your work?

Planning a Journal Paper: Question Sheet

Writing a journal paper is obviously an iterative process. This means that you can't write a paper by going only once through the following questions. However, the questions have been devised to enable you to focus on the important information needed for each section.

OVERALL PRINCIPLE: Think all the time about what the readers of your paper will need.

For the first steps, see ***How to Start Writing a Journal Paper***, page 84.

Stage 1: Title

What will this paper be about? What is its main point?
What information in your title would make your readers read your paper?
Which keywords would you like the title to contain?
Would you like to use abbreviations in the title? If so, are you sure that they are well known?
Would it be appropriate to use a hanging title (a title split by a colon)?

Stage 2: Keywords

Which general keywords do you think you should use?
Which specific keywords do you think you should use?

Stage 3: Illustrations. Remember that the illustrations are often looked at very early in the reading process. They need to be self-explanatory; for each illustration, the title, captions (legends) and labelling need to be fully informative.

What illustrations will you use in this paper?
What do you want them to describe?

Stage 4: Methods

What are the main details of the methods? Think about what the reader *most* needs to know.

Stage 5: Results

What are the main points of your results? Are they shown in the illustrations?

Stage 6: Introduction

Why are you doing this work? Why is it important?
What is your main objective in this paper?

For the beginning of the *Introduction*:
What is the main relevant current knowledge?
For the middle of the *Introduction*:
What is unknown (or a problem with what is known)?
For the end of the *Introduction*:
What is your main objective in this paper? (see above)
What methods will you use to achieve it?

Stage 7: Discussion

Beginning:
How would you VERY BRIEFLY restate your aim?
How would you VERY BRIEFLY summarise the results?
Middle:
How would you place your work in the context of other people's work?
Whose does it support?
Whose does it contradict?
How would you build up the arguments towards your conclusions?
End:
What conclusions can you draw?

Stage 8: Abstract

What points do you think a reader would like to see in your *Abstract*? List them as follows:
The beginning:
What is the context (background) of your work?
What is the gap in the knowledge? Why are you doing the work?
What will be the first sentence of your *Abstract*?
The middle (Part 1): Briefly describe your methods (if necessary).
The middle (Part 2): Briefly describe your results.
The end: Briefly describe your main conclusion(s).
Use signalling words for each of these elements.
Next stage: Be very critical here: Is your Abstract really informative? Or does it contain some descriptive elements?

7 Progress Reports

This chapter covers:

- Reports to an outside organisation or funding body, to show the progress of your research in relation to your original research proposal (see also Chapter 5: *A Research Proposal)*
- Progress reports of a project team
- Checklists

A Progress Report to the Funding Body or Organisation

Purpose

To a funding body: To give a description in academic terms of your progress.

To a commercial organisation:

1. To show the organisation that its money is being appropriately spent, and the organisation will eventually reap benefits from your work (*see Point 1, How to Write It, below*).
2. To give the organisation a report that its members can understand *in their own terms.* (Remember there may be no expert in your field on the organisation's staff (*see Point 2, How to Write It, below*).
3. That your work is progressing along the lines of your original proposal. See also Chapter 5: *A Research Proposal.*
4. That your results are valid and non-trivial.
5. (Possibly) That your work is opening up into other directions *that will be of benefit to the organisation.*

How to write it

A funding body usually has guidelines for progress reports. Follow these, using the principles given in this chapter.

For a commercial organisation, two points need to be kept in mind:

1. **The organisation or company is specifically interested in how your results are going to benefit them.**
Rather than the academic implications of your work, the company will be more interested in how your results will contribute to its competitiveness and profitability. This may not be the case for some large companies, which may be wealthy enough to be able to fund 'blue-skies' research, knowing that in the long term, the academic information will contribute to

Writing for Science and Engineering.
DOI: http://dx.doi.org/10.1016/B978-0-08-098285-4.00007-8

their wealth. However, many companies have to concentrate on their immediate or mid-term business plan. Moreover, if your work is suddenly opening up into an academically interesting and unexpected direction, they may be less enthusiastic than you about pursuing it.

2. **Your report may need to be understood by people with no expertise in your particular field.**
   There may be no one who is familiar with the basic knowledge and terminology of your subject. Even in companies that have the expertise, your report may be passed on to people such as financial personnel.

Your report should therefore:

- Be written so that the organisation can clearly see the benefits to the company's activities.
- Show that you have carried out the work as originally proposed.
- Be written in language that does not need expert knowledge to be understood.
- Use a *Glossary of Terms* to clearly explain terminology that may not be familiar to the organisation.
- Clearly state your proposals for the next stage of the work.

Possible Structure

If you are directing the report to company personnel who are predominantly research-based, it may be appropriate to use the classic *TAIMRAD* structure (*Title, Abstract, Introduction, Methods, Results and Discussion*). If not, the following structure or variations of it may be appropriate.

Section	Cross-Reference to Relevant Part of This Book
Title Page This may need to contain the following features: • Title of the project • The number of the report in the series that you are preparing (e.g. Report number 2) • Date • Suggested wording: Completed for (*name of company*) to fulfil the requirements stipulated in the (*name of the contracting organisation*) contract (*date*) • Your name, department, institution, contact phone and fax numbers, e-mail address	**Title** and **Title Page**, Chapter 2: *The Core Chapter*, page 19
Executive Summary A summary of no more than one page of your main results and main recommendations. Keep in mind the two points in ***How to Write It***, above.	Chapter 3: *An Abstract; a Summary; an Executive Summary*, page 54
Table of Contents	**Table of Contents**, Chapter 2: *The Core Chapter*, page 23

Glossary of Terms and Abbreviations Terms explained so that a non-expert in the field can understand them.	**Glossary of Terms and Abbreviations,** Chapter 2: *The Core Chapter*, page 27
Recommendations	**Recommendations,** Chapter 2: *The Core Chapter*, page 40
Techniques Your techniques may have developed or altered from the ones you originally projected. They need to be described clearly but not necessarily in the detail required for a journal paper. If your procedures have remained the same, you still need to describe them briefly with reference back to the previous reports.	**Materials and Methods,** Chapter 2, page 36
Sections appropriate to the topic, describing the results.	
Conclusions A description of your conclusions from the work to date.	**Conclusions,** Chapter 2: *The Core Chapter: Sections and Elements of a Document*, page 39
Updated project plan and expected time frame for the remaining activities A clear description of the expected remaining stages of your research. If appropriate, show how your plan has developed and possibly altered from your original, projected plan. State your expected time schedule for the following: **1.** The various future tasks (preferably with a Gantt chart) **2.** The schedule of future reports that you will write for the organisation	**Schedule of Tasks/Time Management,** Chapter 2: *The Core Chapter: Sections and Elements of a Document*, page 33
If appropriate: **Requirements for the next stage** A statement of what you expect to need from your funding organisation for the next stage of research.	**Requirements,** Chapter 2: *The Core Chapter: Sections and Elements of a Document*, page 35
References or **Bibliography**	Chapter 15: *Referencing: Text Citations and the List of References*, page 169
Appendices	**Appendices,** Chapter 2: *The Core Chapter*, page 42

Intermediate progress reports should briefly refer back to previous reports. **The final report** should do the following:

- Tie up all the work into a logical story.
- Concentrate particularly on the overall results, conclusions and recommendations.
- *If appropriate*, give guidelines about how the work could be further developed (see **Suggestions for Future Research**, Chapter 2: *The Core Chapter*, page 41).

Checklist for a Progress Report to a Commercial Organisation

- ☐ Does your report clearly show how your work will benefit the company's activities?
- ☐ Does it use language that can be understood by a non-expert in your immediate field?
- ☐ Does it explain terminology that may not be familiar to the company's personnel?
- ☐ Does it clearly show the direction your research is taking?
- ☐ Does it refer back to your previous progress reports?
- ☐ Does it explain your progress since your previous report?
- ☐ Does it describe any procedure that you have used and was not projected in previous reports?
- ☐ Does it show the expected time frame for future activities?
- ☐ *In the final report*:
  - ☐ Does it tie up all your work into a logical story?
  - ☐ Does it clearly describe your recommendations?
  - ☐ Does it show how the work could be further developed?

A Project Team's Progress Reports

Purpose

To report at intervals on the progress of a management project undertaken by several individuals.

Possible structures for a series of progress reports

1. Initial report at the start of the activity

This is likely to be similar to a project proposal, in which you describe how you will approach the task. As with any plan, it will involve intelligent and informed guesswork. Use the principles given in Chapter 5: *A Research Proposal*.

You will probably need the following elements:

	Cross-Reference to Relevant Part of This Book
1. Executive Summary	Chapter 3: *An Abstract, a Summary, an Executive Summary*, page 60
2. Objectives	**Objectives**, Chapter 2: *The Core Chapter*, page 30
3. Initial analysis of the problem	**Problem Statement**, Chapter 2: *The Core Chapter*, page 32
4. A preliminary literature survey	Chapter 4: *A Literature Review*, page 63
5. A clear statement of how you propose to tackle the first stages of the project, together with a brief description of the methods you will use	**Materials and Methods**, Chapter 2: *The Core Chapter*, page 36

6. *If needed*: **Schedule of Tasks** 7. **Allocation of responsibilities** A description of the roles of the various individuals in the team.	**Schedule of Tasks/Time Management,** Chapter 2: *The Core Chapter*, page 33 **Allocation of Responsibilities**, Chapter 2: *The Core Chapter*, page 33

2. **Intermediate reports**
   For intermediate progress reports, use the principles given in this chapter in *Progress Report to the Funding Body or Organisation*, page 111.
3. **The final report**
   Again, use the principles given in *Progress Report to the Funding Body or Organisation*, page 111, and also take the following into consideration:
   - This report will probably need to be longer than the preceding reports.
   - It will need to tie up the whole body of work into a logical story.
   - It should concentrate on the **results**, **conclusions** and **recommendations**.
   - *If required*, include a description and possibly a peer review of the tasks undertaken by the various individuals.
   - *If appropriate*, include a description of how the work you have done could be further developed in the future. See **Suggestions for Further Development**, Chapter 2: *The Core Chapter*, page 41.

Checklist for a Project Team's Progress Reports

Does your first report:

- ☐ Give your team's objectives?
- ☐ Give an initial analysis of the problem?
- ☐ Give a brief preliminary survey of the literature?
- ☐ Describe the various tasks you foresee?
- ☐ Describe the techniques you will use?
- ☐ Give a schedule and time frame for the various tasks?
- ☐ Describe the roles of the various individuals?

Do your intermediate reports:

- ☐ Clearly show the direction your work is taking?
- ☐ Describe your progress since your previous report, with brief reference to those reports?
- ☐ Describe any method or technique that you have used and was not projected in previous reports?
- ☐ Give your conclusions from the work so far?
- ☐ Show the expected time frame for future activities?

Does your final report:

- ☐ Tie up the whole body of work into a logical story?
- ☐ Concentrate on results, conclusions and recommendations?
- ☐ *If required*, give a description of the tasks and peer review of the various individuals in the team?
- ☐ *If appropriate*, show how the work can be developed further?

8 Consulting/Management Report and Recommendation Report

This chapter covers the types of reports for the following purposes:

1. As a report directed to the senior management personnel of an outside organisation, as a result of work you have carried out for the organisation.
2. As an internal assignment for a university course of study. In this case, your assessor will probably need it to be written as though it is for an outside organisation. Check that this is so.
3. Checklists

See also **A Project Team's Progress Reports**, Chapter 7: *Progress Reports*, page 111.

A Consulting *or* Management Report

Purpose

To inform senior management personnel of the results of consulting work you have carried out for them.

> **IMPORTANT: This report should be written so that someone with no technological or scientific knowledge can understand the overview, meaning and implications of it.**

A management report is likely to be read by non-experts

When writing this report, imagine that the academic staff member who will assess it is a senior manager in a commercial organisation. The staff member will, of course, read the whole report and has the background to understand the science or technology behind it. But, in imagining this organisation where your report is going, the following points need to be remembered:

1. **Senior management may not read a whole report. They rely heavily on the *Executive Summary*, *Conclusions* and *Recommendations* to give them an overview of the substance of the report. A longer report may also need section summaries**. They will expect to understand the following from these:
   - What your work means
   - How it will benefit the company's activities

Writing for Science and Engineering.
DOI: http://dx.doi.org/10.1016/B978-0-08-098285-4.00008-X

- Any further work that needs to be done
- (*Probably*): What it will cost

See:
- **Executive Summary** in Chapter 3: *An Abstract, a Summary, an Executive Summary*, page 60
- **Conclusions** (page 39) and **Recommendations** (page 40) in Chapter 2: *The Core Chapter*.

2. **Senior management of an organisation may not have your specific knowledge of the field**. They may have expert scientific or engineering that is not specifically in your area or be even further removed such as accountants or lawyers and so on. Moreover, your report may be passed on to other people whom you may not have expected to read it, e.g. local government personnel. This means the following:
   - **Your report should be able to be understood by non-experts** – at least in overview, meaning and implications.
   - It also needs to be written so that everyone can extract from it what they need, ***without having to read the whole document***.
   - Even if they read the whole document, readers will need a clear pathway to help them navigate through it. The *Executive Summary*, *Recommendations*, *Conclusions* and section summaries are crucial parts of this. See ***The Importance of Overview Information***, Chapter 1: Structuring a Document: Using the Headings Skeleton, page 11.

Possible structure of a management report

Section	*Cross-Reference to Relevant Part of This Book*
Title page	See **Title and Title Page**, Chapter 2: *The Core Chapter*, page 19
Letter of transmittal (Cover letter), if needed	See **Letter of Transmittal**, Chapter 10: *A Formal Letter (Hardcopy or Online)*, page 131.
Executive Summary	See Chapter 3: **An Abstract, a Summary, an Executive Summary,** page 60
Recommendations	See **Recommendations**, Chapter 2: *The Core Chapter*, page 40
Table of Contents	See **Table of Contents**, Chapter 2: *The Core Chapter*, page 23
List of Figures	See **List of Figures**, Chapter 2: *The Core Chapter*, page 26
List of Tables	See **List of Tables**, Chapter 2: *The Core Chapter*, page 26
Glossary of Terms and Abbreviations	See **Glossary of Terms and Abbreviations**, Chapter 2: *The Core Chapter*, page 27
Acknowledgements	See **Acknowledgements,** Chapter 2: *The Core Chapter*, page 23
The following four sections may be required in a consulting report:	
Purpose Statement	See **Purpose Statement**, Chapter 2: *The Core Chapter*, page 31
Problem Statement	See **Problem Statement**, Chapter 2: *The Core Chapter*, page 32

Section	*Cross-Reference to Relevant Part of This Book*
Scope (or Scoping) Statement	See **Scope Statement**, Chapter 2: *The Core Chapter*, page 32
Procedure Statement	See **Procedure Statement**, Chapter 2: *The Core Chapter*, page 32
Background or *Introduction* or both	See **Background and Introduction**, Chapter 2: *The Core Chapter*, page 32
The body of the report I (structured under a series of headings appropriate to your topic) I	
Conclusions (if not placed at the beginning) or *Conclusions and Recommendations*	See **Conclusions** and **Recommendations,** Chapter 2: *The Core Chapter*, page 39
References (if needed)	See Chapter 15: *Referencing*, page 169
Appendices	See **Appendices**, Chapter 2: *The Core Chapter*, page 42

Checklist for a consulting/management report

- ☐ Is it written in a style that enables a reader with no specialist knowledge in this area to understand it?
- ☐ Does it contain a clear road map made of overview information throughout your document: an *Executive Summary, Recommendations* and *Conclusions*, and in a long report, section summaries? These are vital sections to aid a non-specialist's understanding.
- ☐ Will the reader understand how your work will benefit the company?
- ☐ Do you state what needs to be done next?

A Recommendation Report

Purpose

To make a recommendation or a series of recommendations, supported by a reasoned argument, together with appropriate background material, facts and data.

How to write it

- A recommendation report is focused towards the future: it should show the ability to objectively assess a set of conditions and to recommend actions to be taken.
- Recommendations are your subjective opinions about the required course of action, but this doesn't mean you can go into wild flights of fancy.
- No recommendation should come out of the blue; your report should contain adequate supporting information for each recommendation.

Possible structure of a recommendation report

Title Page	See **Title** and **Title Page**, Chapter 2: *The Core Chapter*, page 19
Executive Summary or ***Summary*** or ***Abstract*** Summarise the background material and your investigation.	See An Abstract, a Summary, an Executive Summary, Chapter 2: The Core Chapter, page 60
Recommendations List your recommendations. Or instead, use a section called *Conclusions and Recommendations* and place it at the end of the report (see below).	See **Recommendations**, Chapter 2: *The Core Chapter*, page 40
Contents Page	See **Table of Contents**, Chapter 2: The Core Chapter, page 23
List of Illustrations (if needed)	See **List of Illustrations**, Chapter 2: *The Core Chapter*, page 44
The following four sections may be effective in a recommendation report, either before the *Introduction/Background* or as sections of it.	
Purpose Statement	See **Purpose Statement**, Chapter 2: *The Core Chapter*, page 31
Problem Statement	See **Problem Statement**, Chapter 2: *The Core Chapter*, page 32
Scope (or Scoping) Statement	See **Scope Statement**, Chapter 2: *The Core Chapter*, page 32
Procedure Statement	See **Procedure Statement**, Chapter 2: *The Core Chapter*, page 32
Introduction or ***Background***	See **Introduction** (page 28) and **Background** (page 30), Chapter 2: *The Core Chapter*
Subheadings appropriate to the topic and covering the methods and results.	
Conclusions Note: You may be required to write a section called *Conclusions and Recommendations*. In this case, place it here at the end of the report, and omit the *Recommendations* section after the *Abstract*.	See **Conclusions, Recommendations**, Chapter 2: *The Core Chapter*, page 39
List of References	See Chapter 15: *Referencing*, page 169
Appendices	See **Appendices**, Chapter 2: *The Core Chapter*, page 39

Checklist for a recommendation report

☐ Are your recommendations clearly stated?

☐ Are your reasons for making these recommendations clearly stated and supported by reasoned arguments?

9 Engineering Design Report

Note: There is no accepted model for design documentation. You need to check what is specifically required in your case.

This chapter covers the basic general requirements:

- The purpose of an engineering design report
- Its readership
- The general characteristics of design documentation
- The *Summary*
- Development of a model
- Design calculations
- Checklist

Purpose of a Design Report

Design reports are used to communicate your solution of a design problem, usually to your boss or a colleague.

The design report is a critical component of the design process. An extremely competent or ingenious design solution cannot be communicated by drawings alone; it needs to be supported by comprehensive documentation.

Readership

The report should be written for another person of equal or greater competence than yourself.

General Characteristics of Design Documentation

1. The report should be **self-contained**, except for references to other specific documents (e.g. contracts, drawings, and standards).
2. **Your report must contain all of the information needed for someone to check how you arrived at your recommended solution.** While carrying out your design, you will have used analysis to demonstrate that your design will actually solve the problem. This needs to be clearly set out in your documentation.

Writing for Science and Engineering.
DOI: http://dx.doi.org/10.1016/B978-0-08-098285-4.00009-1

The Workbook

The workbook – either digital or hardcopy – is maintained throughout the process of designing your solution. All of your analysis will therefore be documented in your workbook as well as in your final report.

In the professional engineering workplace, a design workbook can become a legal issue, if for example a design has catastrophically failed. It therefore needs to be meticulously maintained.

Suggested Structure of Design Documentation

1. **Summary**
2. **Development of a Model**
3. **Design Calculations**

The Summary

Purpose of this section: The Summary should state precisely what the report is about and answer the following questions. *To make sure you don't solve the wrong problem, write up the first two before you start the design.*

1. **What problem does the report address?**
   - If the problem was defined in writing (assignment, tender or contract document), just refer to this briefly and accurately rather than restate the whole problem. Your reader will already know what you were supposed to be doing.
2. **What criteria were set for deciding on an adequate solution?**
   - You can make a sensible design recommendation only if you **understand the criteria that are to be used in judging the success of your design**, and obviously you must know this before you start designing. Again, if these were defined in writing, just refer to the original document.
   - Sometimes there are **other constraints** such as national standards that must be met. These should also be stated.
   - **If the criteria were incomplete or contradictory**, for example between the cost and durability of a new product, you need to decide the relative importance of the criteria to be used in making your decision. You need to explain this in a subsequent section.
3. **How did you model the problem?**
   Outline very briefly the factors influencing how you went about your design:
   - The analysis that was needed
   - How many different options were considered
   - The main factors influencing the design

   Note: If you are designing something that is routine, this section would be very short.
4. **What did you conclude?**
   State the following simply:
   - What you concluded. Refer to drawings or other details of your recommended solution.
   - If you considered various options, summarise why you chose your particular solution.

Development of a Model

Purpose of this section: The first step in an engineering design is to be able to conceptualise the problem in a way that allows standard methods of analysis to be used. This section should explain how you went about this.
The form of this section: You should use diagrams and equations as needed, but tie them into a logical presentation using text. Again, for a routine design, this section need not be very long.

This section would typically answer the following questions:

1. **What assumptions were needed?**
   - **Every analysis of a real system has some assumptions built into it** because the physical world does not behave in the way that engineers need to assume.
   - **Often these assumptions are taken for granted**.
   - **But sometimes you need to make unproven assumptions** to simplify the problem enough to be able to model it for analysis. This is acceptable provided that you **state what the assumptions are and that you check them later**.
   - **Your assumptions should therefore be stated clearly at the start of this section.**
2. **How was the problem modelled?**
   - You should now be able to represent the object or system by a simple conceptual model that is capable of being analysed. Use a diagram (e.g. stick, block, circuit and flow diagram) to show this and discuss it if needed.
3. **What analysis was used?**
   - **State the laws that you have applied.** You should state the relevant physical or other laws that you needed to apply. State them by name; you don't need to write them out or include any proofs.
   - **These laws will probably result in equations.** These should be written out in full, using well-defined variables (include a named list or label to your conceptual sketch).
   - **Simply state the method of analysis you used and make a reference to it.** Nearly always, the method of analysis used is standard and can be found in textbooks.
   - **If the analysis is repetitive** (because the solution has many components of the same type), you need to document all of this only once.

Design Calculations

Purpose of this section: This part proves that your design will work as it should; the section will consist mostly of small sketches and steps in solving equations.
Form of this section: Use subheadings to make it clear what each section is about and emphasise the important results, e.g. by underlining or boldfacing text.

- **Add numbers to your design** and then use conceptual modelling to show that the design will function as it should – i.e. that it will meet the design criteria stated at the beginning.
- **The numbers needed are those that would enable someone to actually make your design**. Include details of all components (material and dimensions of parts, electrical components and so on) plus all the physical properties you have used (e.g. strengths, elastic modulus, density, specific heats).

- **Your design report should contain only your final recommended solution**. In your workbooks, you may have needed to guess some of the numbers to carry out the analysis. If it subsequently turned out that your design did not meet the criteria for success, then you would have changed the guesses and tried again. You may also have tried out several different design options before finding one that worked. No matter how long they took you, **the details of the designs that did not work are irrelevant**. If they should be needed later, they can all be found in your workbook.
- **Where you looked at several very different design solutions**, you may want to include detailed results from the best of each to justify your final choice.
- **For repetitive designs, you may want to use a spreadsheet**. In that case, document one example calculation right through as above, and just show the results of the remaining components on a table (which must include enough of the intermediate results that it can be checked easily).

Checklist for a design report

- ☐ Is it written for another person of equal or greater competence than yourself?
- ☐ Is the report self-contained (except for references to such documents as standards, textbooks, contracts and so on)?
- ☐ Is all your analysis also contained in your workbook?
- ☐ Does it state the design problem?
- ☐ Does it describe the design criteria?
- ☐ Does it describe how you modelled the problem so that standard methods of analysis could be used?
- ☐ Does it state the assumptions you made?
- ☐ Do you tie up diagrams and equations with explanatory text into a logical presentation?
- ☐ Does it state the laws you applied and the method of analysis you used?
- ☐ Do you show that your design meets the design criteria?
- ☐ Would the numbers on your design enable someone to make it?
- ☐ Does it state your conclusions?
- ☐ Do you present only your final recommended solution?
- ☐ If you considered various design options, does it state why you chose your particular solution?

10 A Formal Letter (Hardcopy or Online)

This chapter covers the requirements for hardcopy formal letters that accompany the transmission of large documents, online formal communications and letters of application for a position. All formal letter types have the same requirements.

- The parts of a formal letter
- Overall layout
- Structure of the information
- Style for letter writing
- Various types of letters:
  - Covering letter
  - Letter of transmittal
  - Letter of application
- Checklists

The components of a formal letter need to be arranged in a particular sequence, as dictated by the conventions of formal letter writing.

The Parts of a Formal Letter

All parts are left justified, except for the subject heading.

1. **Your address or institution's address (or letterhead)**
   2-line space
2. **The date**
   2-line space
3. **Name and address of the person you are writing to**
   3-line space
4. **The greeting (salutation)**
   2-line space
5. **The subject heading**
   2-line space
6. **The body of the letter**
   2-line space
7. **The closing**
   Leave a 6/8-line space for your signature

Writing for Science and Engineering.
DOI: http://dx.doi.org/10.1016/B978-0-08-098285-4.00010-8

8. Your written signature with your typed name and position below it
2-line space
9. The letters *Enc.* if you are enclosing additional documentation with the letter

The parts listed above are described in more detail here:

1. **Your address or institution's address** (or letterhead)
   - Left justified. (Note: the conventions of some years ago dictated that it should be right justified. This is now regarded as old-fashioned.)
   - If you are using letterhead paper, an address isn't needed.
   - It is now no longer the convention to put a comma at the end of each line.
2. **The date**

   Use the format: Day (*in figures*) Month (*written out*) Year (*in figures*). No commas.

   Correct: 8 October 20xx

   Incorrect: 8/10/xx (different countries use different formats when using only figures; it can cause confusion); 8th October 20xx (is going out of fashion).
3. **Name and mailing address of the person you are writing to**
   - Left justified.
   - Commas not needed at the end of each line.
4. **The greeting (salutation)**

   According to the tone of the letter, choose from the following:

   Dear Sir or **Dear Madam** (used in formal situations, *either* when you don't know the family name of the person *or* when it would be inappropriate to use it).

   Dear Sir/Madam (used in formal situations when you don't know the family name or the gender of the person you are writing to).

   Dear Mr *surname*; **Dear Ms** *surname*; **Dear Mrs** *surname* (used infrequently in a formal situation and only when the woman has signed herself that way); Dear **Dr** *surname* or **Dear Prof** *surname.*

   Dear (*first name*) used when you are on familiar terms with the person you are writing to but still need to use a formal letter format.
5. **The subject heading (title)**
   - A concise title, two lines below the greeting, centred and boldfaced for emphasis. It should give the reader instant access to the main point of the letter.
   - Don't use *Re:* before the title. It's meaningless and out of date.
   - Don't underline, which is outdated. Use boldfacing instead.
6. **The body of the letter** (see below: *Structuring the Information*, page 130)
7. **The closing**
   - Classical letter-writing conventions dictate the following:

     If you have used **Dear Sir, Dear Madam, Dear Sir/Madam**:
     You must use **Yours faithfully** as the closing.

     If you have used the surname in the salutation:
     You must use **Yours sincerely**.
   - This rigid convention has now been considerably relaxed. Many companies now favour **Yours sincerely**, whatever the initial salutation.

   If the letter is not strictly formal, the tone of the letter can be softened by using **Regards, Kind regards** or **Best wishes**, either before the closing or alone.
8. **Your written signature. Below it, include your typed name and position.**
   - After the closing, leave about eight blank lines for your written signature.
   - Then enter your name (left justified). Use your full first name and surname (e.g. *Joe Bloggs*), not initials and surname (not *J. F. Bloggs*). This conveys to the reader your gender – making you easier to contact – and the sense of a real person behind the letter.
   - On a new line, add your position.
9. **Insert *Enc.* if you are enclosing additional documentation with the letter.**

Overall Layout

- Choose a simple serif (e.g. Times Roman) or sans-serif (e.g. Arial) font. Elaborate fonts are more difficult to read and give the wrong impression.
- 10- or 12-point font.
- In the text of the letter, use single-line spacing and one blank line between paragraphs.
- Use ample margins.
- If the letter is short, adjust the various spacing so that it isn't squashed into the top part of the page.
- The last page shouldn't contain just the signature. If necessary, slightly reduce the font size and/or margins.
- All of the elements should make a pleasing arrangement on the page (see Page 129).

Checklist: the parts of the letter

☐ All left justified, except for the subject heading
☐ Date format: 8 October 2001
☐ Name and address of person you are writing to:
Left justified with no commas at the end of each line
☐ The Salutation:
Dear Sir, Dear Madam, Dear Sir/Madam, Dear Mr …/Ms…/Mrs …/Dr …/Prof…
☐ Subject heading:
Describes the main point of the letter
Centred, boldfaced
☐ The closing:
Yours sincerely or *Yours faithfully*,
Your signature
Your typed name, including your first name; not just your initials and surname
Your position
☐ *Enc.* (if you are enclosing something)

Checklist: overall layout

☐ Simple font, 10 or 12 point
☐ Single-line spacing
☐ One blank line between paragraphs
☐ Pleasing arrangement on the page

Example: Layout of a Formal Letter

Composites Laboratory
School of Engineering
University of Middletown
PO Box 123
Middletown

8 October 20xx

Dr Lesley Green
Director, Research and Development
Composites Construction Ltd
Middletown.

Dear Dr Green,

Research Seminar to the Board of Directors

Body of the letter

Yours sincerely,

Pat Black (Dr)
Isaac Newton Research Fellow, Composites Development

Structure of the Information

Overall principle
Place the main point at the beginning and the supporting information after the main point.

- **First paragraph**: (*if appropriate*) A courteous acknowledgement of the letter/phone call and so on.
- **Second paragraph**: The main point.
  - **This is important.** It applies to all letters, including those conveying bad news.
  - Don't build up to the main point. The main point should be at the beginning with the supporting information below it. (*Rule of thumb*: This is not a detective story. Don't lead up to the disclosure at the end; start with it.)
- **Last paragraph**:

Don't finish abruptly. A courteous final paragraph is needed, e.g. *I am looking forward to your response.*

Checklist: structure of the information

- ☐ (*If appropriate*) Does the first paragraph courteously acknowledge their letter/phone call?
- ☐ Is the main point at the beginning of the letter?
- ☐ Does the supporting information come after the main point and not lead up to it?
- ☐ Have you avoided finishing abruptly?
- ☐ Does your final paragraph make a courteous finish?

Style of Writing

Overall principle
Write as you would speak in a comfortable, serious conversation.

1. Use plain language. Write to inform, not to impress.
2. Put yourself in the reader's mind, and work out how he/she would react to your language.
3. Avoid the old-fashioned, stuffy phrases associated with classic formal letter writing.
4. Make sure the spelling and grammar are correct.

The style directives listed above is described in more detail here:

1. **Use plain language; write as you would speak in a comfortable, serious conversation.**
   Imagine that you are across the table from the person to whom you are writing, or on the telephone. Write in the way you'd speak in these situations but without colloquialisms or contractions (*don't, wouldn't*, etc.; see *Contractions*, Chapter 18: *Problems of Style*, page 208).
2. **Put yourself in the reader's mind, and work out how he or she would react to your language.**
   It is possible to innocently write something that could be interpreted quite differently by the reader.

For this reason, **stand away from your personal involvement in what you have written, and try to interpret it in the way the reader may see it.** It's not easy, but it's absolutely necessary.

3. **Avoid the old-fashioned, stuffy phrases associated with classic formal letter writing. Express the idea in plain English.**

Do not use phrases such as:	Use these instead:
As per	In accordance with
Attached hereto *or* herewith	I am *or* We are attaching *or* Attached is…
Enclosed hereto *or* herewith	I am *or* We are enclosing *or* Enclosed is…
Pursuant to your request	Following your request…
We are in receipt of your letter	Thank you for your letter
We are pleased to advise *or* I am pleased to advise	We are pleased *or* I am pleased to tell you that/let you know that…
You are hereby advised	This letter is to let you know that…
Please contact the writer	Please contact me

4. **Make sure the spelling and grammar are correct**
   - Spell check at the very end of writing.
   - But proofread it thoroughly afterwards once again. The spell-checker can pass words that you did not mean (e.g. *as* instead of *at*, *hit* instead of *him*; see Chapter 18: *Problems of Style*, page 212).
   - If you know your grammar may be faulty, give it to someone to check.

Checklist: style of writing

- ☐ Have you used plain language?
- ☐ Is your letter easy to understand for someone without your level of knowledge?
- ☐ Have you made the letter sound personal?
- ☐ Have you checked it to see if your phrasing could be misinterpreted?
- ☐ Have you avoided using the classic, stuffy letter-writing phrases?
- ☐ Have you spell-checked the *absolutely final version*?
- ☐ Are you sure that it is grammatically correct?

Types of Formal Letters

Letters that accompany a document

1. A ***Covering Letter*** is any letter that is sent together with any document.
   Purpose of a *Covering Letter*
   - To provide the recipient with a specific context within which to place the document
   - To give the sender a permanent record of having sent the material
   - To show willingness to provide further information
2. A ***Letter of Transmittal*** accompanies formal documents such as reports or proposals.
   Purpose of a *Letter of Transmittal*
   To achieve one or more of the following:
   - To identify the report topic, and scope or extent of the study
   - To give an overview outlining the main aspects of the primary document

- To identify the person who authorised the report and the date of authorisation
- To call for a decision or other follow-up action
- To emphasise any particular points you may want to make
- To show willingness to provide further information

3. **Structure of a *Letter of Transmittal***

It should be **brief**.

- **First paragraph**: Describe what is being sent and the purpose of sending it.
- **Middle section**: A longer letter may summarise key elements of the proposal and provide other useful information.
- **Final paragraph**: Establish goodwill by thanking the recipient and showing willingness to provide further information.

Letters of application

Purpose

To convince a prospective employer that you are a worthwhile candidate for an advertised position. This type of letter accompanies your CV (résumé).

How to Write It (Use the schematic on page 133 for the conventions of wording and formatting.)

Since it is your first approach to your prospective employer, the letter should be polished and professional, free of mistakes and well formatted.

It should contain the following information:

1. **In the first paragraph:** state the specific position that you are applying for, with the job title and the vacancy number if there is one. State where the position was advertised.
   Example:
   I would like to apply for the position of… (*Job title, Reference number*) advertised in… (*Source*) on Thursday 8 October.
2. **In the second paragraph:** Your qualifications and experience that are particularly relevant to the position.
   *Exam*ple:
   I have a master's degree in Resource Management (Upper Second Class Honours) from the School of Environmental Science, University of Middletown, specialising in water quality.
3. **In the third paragraph (a fourth paragraph may also be needed):**
   - **Refer to your CV.**
   - **State why you are interested in the position and in the specific organisation**. This can include work experience and your aspirations.
   - **State what you have to offer the organisation**. Relate your academic record and work experience to your knowledge of the activities of the organisation, stating the relevant skills and abilities you believe you can bring.

     This requires some previous research into what the organisation does. Human resources personnel often state that they take as much – sometimes more – notice of the covering letter than of the CV; they look for evidence that the applicant has done some homework and thought about her or his skills relevant to the organisation's activities.
4. **In the final paragraph, ask for an interview**, and show your willingness to expand on the information contained in the CV and letter. Include information on your availability and

where you can be contacted (preferably a phone number where messages can be left, an e-mail address or a fax number).

Guidelines on style

Don't be too modest or hesitant. You need to sound enthusiastic and confident, striking a balance between that and sounding bumptious.

Do not be afraid to use *I*, together with suitable verbs to describe your achievements. *For example: I have developed…, I have initiated…, I managed…*

Use clear, direct language. Don't try to impress with long sentences and big words.

Example of a letter of application

	Your address
	Date
	Recipient's address
	Dear Ms. White,
	Vacancy No. AF/34: Environmental Planner, EnviroCorp
Paragraph 1: Specific position/job title/vacancy number/source.	I would like to apply for the position of Environmental Planner within the Environmental Planning Section of Peterson Associates Ltd (Vacancy Number AF/34), which was advertised in *The Independent* on Tuesday 14 March 2000.
Paragraph 2: Qualification and experience relevant to the position. Show knowledge of the organization.	I have a PhD in Environmental Science from the University of Middletown; the thesis topic was 'Biofilm Development in a Subsurface Flow Wastewater Treatment Wetland.' I am very interested in developing my career as an environmental planner, in particular with an interdisciplinary team in an organization with an international reputation for water quality issues such as EnviroCorp. I found your recent work on the Wylie Stream intake particularly interesting because of the difficult nature of the associated water quality issues.
Paragraphs 3 and 4: Refer to CV. Your goal. Relate goal and qualifications to the organization.	As my enclosed CV shows, my PhD topic and activities over the last six years have been directed towards my goal of becoming an environmental planner. In addition to the experimental work, my PhD also required me to be very conversant with water quality legislation. This was also needed for the three-months project for the Farleigh Community Board, which resulted in a detailed written report and a presentation to three community groups. I expect to be able to bring my skills in data analysis and my understanding of legislative procedures to the position of environmental planner, and hope to expand them considerably.
	In addition to a good academic record, I also have wide-ranging interests and I enjoy working with people, particularly with community groups. My communication skills are well developed: I have made several presentations on water quality issues to community groups and also written several reports.

Last-but-one paragraph: Ask for interview. Information on your availability. Contact information.	I would very much like an opportunity to discuss my application more fully with you. I am available for interview at any time that is convenient for you. My telephone number is 012 453–6824; messages can also be left for me on 014 584–8834. I can also be reached by fax, number 014 473–8356, or by email: *j.brown@freenet.net*
Final paragraph: courteous closing.	I very much look forward to hearing from you. Yours sincerely, J Brown Jane Brown (Dr) Enc.

11 Emails and Faxes

This chapter covers the transmission of formal, professional material by email and fax.

Formal Emails to Communicate Work Matters

A formal email should be written as though it is a business letter. Moreover, because of the immediate nature of both receiving and sending emails, some organisations have developed policies of etiquette for them. When using emails to communicate work matters, you need to take the following into account:

Style of writing	**1.** Take as much care writing a formal email as you would in writing a letter. Be careful what you say and how you say it. **2.** Don't use the pop conventions of the email culture. Lowercase letters at the start of sentences, *i* instead of *I* and *u* instead of *you* will make a poor impression. **3.** For people you know, it may be appropriate to start the message with the person's name followed by a colon. **4.** For someone you don't know or are on formal terms with, start with the conventional Dear Mr/Ms/Mrs/Dr/Prof [see Chapter 10: *A Formal Letter (Hardcopy or Online)*]. Close with the corresponding closing. **5.** Structure the content of your message in the same way as you would a letter. Don't do a brain dump.
Confidentiality	Don't assume that an email is confidential. Never put in a mail message something that you wouldn't want other people to read. Some people call an email a 'postcard to the world'.
Permanence	Don't regard your files – sent or received – as in safe keeping. Networks are not fail-safe. Make sure that you store hard or digital copies of anything important.
Commercial sensitivity	No commercially sensitive material should be sent by email.
Contractual material	Avoid using email for contractual material unless it is followed by hard copies.

Writing for Science and Engineering.
DOI: http://dx.doi.org/10.1016/B978-0-08-098285-4.00011-X

Attachments	**1.** When sending or receiving attachments, make sure they are free of viruses. **2.** Check the size of file attachments before you send them. If they are large, zip the file first; this avoids transmission-decoding problems. **3.** Any files sent via email must have the permission of the author.
Forwarding unnecessary messages	Don't send unnecessary messages, particularly when forwarding material to large groups. The minor effort involved is far outweighed by the irritation it can cause.
Content of auto-signature	Make sure your auto-signature contains your name, address of your institution and telephone and fax numbers. You may also want to include such things as a website URL. If your system doesn't carry an auto-signature function, then make up a template, and use that for each message.
Angry or inflammatory remarks	Don't write a formal email as a fast reaction when you are feeling irritated with the recipient. You'll probably regret it.

Formal Faxes

As with emails, take as much care as you would when writing a letter.

1. If possible, use a fax template for the cover document. This will lay out all of the necessary material such as the recipient's name, institution, fax number and so on.
2. **If you are faxing to someone you don't know or are on formal terms with:**
   - Use the conventions for starting and finishing letters [see Chapter 10: *A Formal Letter (Hardcopy and Online)*, page 125].
   - Structure the content of the fax as you would a letter. Don't do a brain dump.

Checklist for formal emails

- ☐ Have you used the conventions for the salutation and the closing of a letter?
- ☐ Is the content of the email structured as you would a letter?
- ☐ Have you avoided sending commercially sensitive or contractual material by email?
- ☐ Have you made hard copies of important emails, both sent and received?
- ☐ Have you scanned attachments for viruses?
- ☐ Have you zipped large files if they are to be sent as attachments?
- ☐ Do you avoid forwarding unnecessary messages?
- ☐ Does the auto-signature contain your name, address, telephone number and fax number?
- ☐ Have you written it when you are feeling irritated with the recipient?

Checklist for formal faxes

- ☐ Have you, if possible, used a fax template for the cover document?
- ☐ Does the fax use the conventions for starting and finishing letters?
- ☐ Is it structured like a letter?

12 A Procedure or a Set of Instructions

This chapter covers:

- The structure of an efficient set of instructions
- The required language
- Checklist

Purpose

To explain to someone how to do something clearly, precisely and accurately. Procedures may provide the following:

- Steps for assembling something
- Steps for operating something
- Steps for maintaining, adjusting, repairing or troubleshooting something

Difficulties

Knowing who you are writing for and under what circumstances they will be using the product or system you are describing. This type of 'how-to' writing is often confusing and poorly written. It comes from not realising that most other people do not have the in-depth knowledge of the system that you do.

To avoid this, place yourself in the reader's mind and work out what they need to hear from you. It can be difficult to distance yourself from your own knowledge and do this.

How to Write it

This section is divided into two parts:

1. Possible structure for a procedure
2. Guidelines for wording the instructions

Writing for Science and Engineering.
DOI: http://dx.doi.org/10.1016/B978-0-08-098285-4.00012-1

Possible Structure for a Procedure

Section	Cross-Reference to the Relevant Part of the Book

1. Introduction
Include a short purpose statement. This is important. Many writers assume that because they are telling readers what to do, there is no point in telling them why to do it. This is dangerously wrong; most people won't automatically see the wisdom of doing it your way or the possible dangers involved. Include the following:
• Introduce the material, explaining the **purpose** for and the **importance** of the instructions.
• Give a brief overview to help the reader understand:
  • How the product or system works
  • Why the instructions must be followed
  • What will be achieved
	See **Introduction**, Chapter 2: *The Core Chapter*, page 28

2. Glossary of Terms and Abbreviations Give clear, precise definitions. Place this section at the front of the document so that the reader can easily find them.	See **Glossary of Terms and Abbreviations**, Chapter 2: *The Core Chapter*, page 27
3. Tools and materials required List the tools and the materials that the reader will need.	
4. Special instructions, e.g. safety warnings Prominently display any special items such as safety warnings. But make sure that an important warning is also repeated in the instruction to which it relates.	
5. **Then give a numbered series of instructions** (for guidelines, see below).	

Guidelines for the Wording of the Instructions

Headings that have widely accepted meanings

DANGER	Reserved for steps in a procedure that could lead to injury or loss of life.
WARNING	Used for steps that could result in damage to the product.
CAUTION	Used where faulty results could occur.
COMMENT	Used to: • Alert the reader to a potential problem. • Make suggestions that would make the reader's task easier.

Summary of guidelines

1. Remember that most people do not read a complete set of instructions before they start.
2. Use the imperative form of the verb, i.e. one that gives an instruction.
3. Use the pattern *if…then*. Do not give the instruction first.
4. If there is a safety aspect, give the warning first. Do not give the instruction first.
5. Do not leave out vital information.
6. Do not leave important actions to the discretion of the reader.
7. Be clear and unambiguous.
8. Let each instruction require only one action.
9. Use simple words.
10. Write one-way directions.
11. Be safety-conscious.

The guidelines summarised above are described in more detail and with examples below:

1. **Remember that most people do not read a complete set of instructions before they start.** When you write, remember that the reader is carrying out each of your instructions while reading it, without knowing what comes next. Two things to remember:
   a. **Write chronologically**, i.e. take into account the sequence in which people will do things.

Poor Example
A set of assembly instructions for a piece of kitset furniture, which has as its final direction: *Remember to glue all pieces as you assemble them.*

   b. **If the procedures are conditional on something (*if…*), say so at the beginning.**

Poor Example	*Rewritten as the First Instruction of a Set*
A procedure in which the final instruction is: *If the temperature is above 18°C, DO NOT carry out the above procedures.*	**If the temperature is above 18°C, DO NOT carry out the following procedures.**

2. **Use the imperative form of the verb, i.e. one that gives an instruction.**
   Give orders clearly so that there is no mistaking what you mean. Avoid the word *should.*

Poor Examples	*Rewritten as an Instruction*
The power switch should be turned off. Or *You should turn the power switch off.*	**Turn the power switch to OFF.** Negative instructions are also effective. **Do NOT turn the activator dial.**

3. Use the pattern *if… then*. Do not give the instruction first.

The pattern 'if…then' asks the reader to consider whether the condition applies before carrying out the action.

Poor Example	*Rewritten, Stating the Condition First*
Push the red button, but only if procedure A has failed.	**If procedure A fails, push the red button.**

4. If there is a safety aspect, give the warning first. Do not give the instruction first.

Poor Example	*Rewritten, Giving the Warning First*
Light the match and slowly bring it towards the nozzle. Do not light the match directly over the nozzle.	**WARNING: Do not light the match directly over the nozzle.** **Light the match and slowly bring it towards the nozzle.**

5. Do not leave out vital information.

Remember that you are familiar with the procedure; your reader is not. Do not assume that the reader will understand what you meant to say. State it explicitly so that the reader does not have to think, only to act.

6. Do not leave important actions to the discretion of the reader.

Avoid words such as *should* and *may*.

Poor Example	*Rewritten So That Reader Does Not Have to Use Discretion*
The condensate line may need to be drained.	**1. Read and record the condensate level on the sight glass.** **2. If level is greater than 15 cm, open Valve D.** **3. Drain the condensate line.** **4. Close the valve when steam begins to come out of the valve.**

7. Be clear and unambiguous.

- Write your instructions from the position – quite literally – of the reader.
- Terms such as *front* and *back*, and *left* and *right* can be confusing. If you were standing in front of the device when writing the instructions, left and right are reversed for the repair technician standing at the back. And it might be the right way up when installed but upside down when being repaired.
- Therefore, avoid these terms whenever possible, or carefully explain the viewing direction.

Poor Example	*Rewritten from the Reader's Position*
Make sure the switch is in the upwards position, and then close the drain valve.	**Make sure the switch is in the OFF position, then close the drain valve.**

8. Let each instruction require only one action.
The possibility for confusion is reduced if you make sure that only one action is contained in each numbered instruction.

Poor Example	*Rewritten as a Series of Actions*
Ensure that both the water supply valve and the feed valve are open, and then start the slurry transfer pump by pressing the START button.	**1. Ensure that both the water supply valve and the feed valve are open.** **2. Then start the slurry transfer pump by pressing the START button.**

9. Write one-way directions.

Poor Example	*Rewritten as One-Way Directions*
These instructions are in the reverse order. Obeying them in the sequence given could ruin the pump and probably damage the preset control valve.	
1. *Start the pump.*	**1. Make sure that the control valve is open. Do not adjust it; it is preset.**
2. *Before starting the pump, check to see that the cooling-water valves are open and that the control valve is open.*	**2. Make sure that the cooling-water valves are open.**
3. *The control valve is preset and should not be adjusted.*	**3. Start the pump.**

10. Use simple words.
Use the simplest words possible.
Do not use jargon, e.g. don't say *deactivate* when you mean *turn it off.*

11. Be safety-conscious.
Safety instructions and warnings must be well thought out and prominently placed at the beginning of the instructions.

Poor Examples	*Rewritten*
The main point of this message – that there is a danger of flashback – is at the end.	The main message is now at the beginning, highlighted by the word **DANGER**. This is followed by a strong negative instruction.
• *Valve X is not to be opened before cooling to 18°C because of the possibility of flashback.* • *Do not open Valve X before it has cooled to 18°C; there is a danger of flashback.* • *On no account should you open Valve X before it cools to 18°C, or you may cause flashback.*	**DANGER OF FLASHBACK: DO NOT open Valve X before it cools to 18°C.**

Common Mistakes

Procedures that lack detail or are confusing. This arises from the following:

1. Not taking into account that your reader will know far less about the system or product than you.
2. Using complex or ambiguous language.

Checklist for a procedure or set of instructions

- ☐ Do you use the words **DANGER, WARNING, CAUTION** and **COMMENT** in accordance with their widely accepted meanings?
- ☐ Do you list the tools and materials required?
- ☐ Is there a *Glossary of Terms and Abbreviations* that gives clear, precise definitions?
- ☐ Are special instructions such as safety warnings prominently displayed?
- ☐ Is the imperative form of the verb used?
- ☐ If the instruction is conditional on something, is the pattern *if... then* used?
- ☐ If there is a safety aspect, is the warning given first, before the instruction?
- ☐ Has any vital information been left out?
- ☐ Have any important actions been left to the discretion of the reader?
- ☐ Is each instruction clear and unambiguous?
- ☐ Does each instruction require only one action?
- ☐ Is each direction one-way?
- ☐ Is the wording simple? Have you avoided jargon?
- ☐ Is the whole procedure safety-conscious?

13 Thesis

This chapter covers:

- The purpose of a thesis
- Difficulties of writing a thesis
- Writing up as a process to be managed
  - To write up as you go along or at the end?
  - Stages of the final write-up
- Structure of a thesis
  - Elements likely to be needed
  - Other possible useful sections for a thesis

Note: Much of the material in Chapter 6, *A Journal Paper*, is also relevant to the writing of a thesis.

Purpose of a Thesis

To show to a very small number of expert assessors (probably fewer than five) your competence in pursuing and writing up a body of independent research.

Implications of this:

- Your writing should be aimed at a level appropriate to experts.
- The main material of your work will be – or may have already been – written up as conference or journal papers. The papers will represent a concentration of the work in the thesis.
- The thesis therefore needs to describe all of the work you have done, without being a blow-by-blow account of every piece of data you collected.

Difficulties of Writing a Thesis

1. **The sheer size of a thesis**
   One essential difference between a thesis and any other piece of graduate writing is that of size. Because of this, a thesis is for many people a worrisome event at the end of the experimental work. Very few graduates enjoy writing up their work more than actually doing it; to many, writing up can be a stressful process. It can also take very much longer than expected.

Writing for Science and Engineering.
DOI: http://dx.doi.org/10.1016/B978-0-08-098285-4.00013-3

For these reasons, **leave more time to write up than you would expect it to take, especially for a Ph.D.**

2. **File handling**
   Because of the ultimate size of the document, it is worthwhile consulting an IT specialist for information about file handling, importing graphics and so on.
3. **The need for backups**
   This can't be stressed enough.
   In every institution, there will be stories about disasters occurring from files not being backed up. In spite of this, it is still remarkable how casual people can be about adequate backup. Ideally, keep backups at different locations (e.g. work and home) and update them regularly – at least once per week.
   An effective way of scaring yourself into backing up fully and frequently is to make a mental list of all the ways you could lose your files:
   - Burglary of your home or institution
   - Theft of your car
   - Viruses
   - Collapse of the hard drive
   - USB failure or loss
   - Power supply failures
   - Errors by other people using the same computer and so on.

How to Write it: Writing Up as a Process to Be Managed

To write up as you go along or at the end?

Supervisors often try to encourage students to write up while doing the experimental work. Is this a good thing to do?

Advantages of writing up as you go along:

1. You are writing something up while it is still fresh in your mind. This shouldn't be underestimated. It is very easy to forget after a year or so the details of a procedure that was once second nature to you.
2. It lessens the burden of a massive piece of writing at the end.

Advantages of leaving it until the end:

1. It is often difficult enough to keep up the momentum of the experimental work, without having to deal with writing as well.
2. If you hate writing, it allows you to put it off.
3. Students have commented that it is a waste of effort if you don't know what you're talking about at that particular time. This is an instance of gaps in your knowledge, not of muddled thought; understanding of a topic and its implications usually increases with time.

4. **Keeping records.** If you do leave it until the end – as many do – it is essential that you keep exceptionally comprehensive records of your work in progress. It is horrifying what you think you'll never forget, and then do so.

Conference or journal papers

If you have already written up papers from your work, you'll find that this will help to tighten ideas about how to process your thesis.

Stages of the final write-up

1. **Keep in mind what you've done, how you've done it and what's new about your research.**
2. **Work out a basic structure for your thesis.**
There will be several ways in which your work can be structured. You need to work out the optimal way to present *your* material. It may be quite different from that of other people in your work group. Given the mass of information that you are likely to have by the end of your experimental period, deciding on an appropriate structure can be a problem. Many people find that while they can't work it out, they can describe it adequately to a friend; this person could take notes for you or use a voice recorder. See *Structure of a Thesis*, this chapter, page 147.

Using the *Outline mode and Master Document mode of Microsoft Word®*

It is worth becoming familiar with the *Outline* mode of Microsoft Word®. This mode will help in the initial organisation, revising and editing of your document. See **The Outline Mode of Microsoft Word®: Organising a Document**, Chapter 1: *Structuring a Document: Using the Headings Skeleton*, page 11.

You can also use the *Outline* mode in conjunction with the *Master Document* mode. A master document is a 'container' for a set of separate files (or subdocuments). You can use a master document to set up and manage a multipart document, such as a thesis with several chapters. For example, you can view, reorganise, format, proof, print and automatically create a *Table of Contents* for multiple documents treated as a whole.

3. **Draw up a preliminary outline of headings and subheadings.**
It will look like a *Table of Contents* without the page numbers. See *Table of Contents*, Chapter 2: *The Core Chapter*, page 23.

Work out your system of headings and subheadings. But keep it flexible; it's very much an iterative process, and you'll need to keep changing it as your ideas evolve. The worst way to work with an outline is to try to shoehorn your ideas and results into an initial rigid structure.

Your initial *Table of Contents* may not bear much resemblance to your final one. You can only do it as far as you can see. The final *Table of Contents* and the finalised structure may take its final form only very late in the write-up.

4. **Write the easiest parts first.**
This is usually anything to do with experimental procedures (see **Materials and Methods**, Chapter 6: *A Journal Paper*, page 93, and Chapter 2: *The Core Chapter*, page 36).

If you are dealing with complex mathematical solutions, you may find that writing the material for your Appendices, where lots of derivations may be needed, can help in clarifying your thoughts.

5. **Other late stage tasks.**
   Graduates say that most of their time is spent working out how to analyse and present their data optimally. Some of the tasks involved are:
   a. Thinking about what the data means.
   b. Considering how it relates to the published literature.
   c. Deciding how to discuss it accurately and succinctly.
   (You'll probably find that you can't think about these three things for more than a few hours at a time.)
   d. Reading the current literature.
   e. Analysing the data and presentation:
   - Everything in a science or engineering thesis hinges on how you analyse and present your data.
   - If you have large spreadsheet files, it may take days on each file before you can finally work out how to present the data optimally.
   - After you know what your data are saying, the linking text should readily follow.

   f. Spending a great deal of time on the final formatting.
   g. Rewriting, losing the file(s) and so on.
6. **Writing the literature review.**
   The literature review needs to be written as one of the final stages of the process because your understanding of the interconnections within the literature and of your work will increase with time. It also needs to incorporate the relevant literature that appears immediately before submission of your thesis.

   For these reasons, the initial literature review that you may have written at the beginning of your research will be inadequate for your thesis. See Chapter 4: *A Literature Review*, page 63.
7. **For referencing: Tables can be an effective way of presenting large amounts of material.**
   Tables can be very useful in the following, e.g.
   - A section called ***Review of Methods***, particularly for mathematical work. The references can be effectively tabulated, so that the various mathematical methods of solution are displayed alongside their author(s).
   - The ***Literature Review*** or ***State of Knowledge***. As an addition to the text, it may be worth considering using a tabulated presentation to summarise the content of each of the relevant papers.

Date	Author(s)	Title	Comments	Reference Number
Listed chronologically			Your comments on the content of the paper.	If the numbering system of referencing is being used (see page 171), the unique number is used in the text and the *List of References*.

Various headings in the list of the cited papers could be used. For example, the general review books and papers could be listed first, followed by other sections appropriate to the topic, with papers listed chronologically in each section.

8. **Revising and proofreading the thesis.**
   For the guidelines for revising and proofreading, which are both essential processes to ensure a professional document, see Chapter 17: *Revising*, page 169.
9. **Formatting for appearance.**
   Your institution will have specific regulations on aspects of formatting.

   If your thesis topic is one where there are a number of related previous theses, then you have a good range for getting ideas of formatting, style, requirements and so on. But be cautious if there are only one or two: you won't know whether they are good models. Errors tend to be propagated in this way.

   Specialist textbooks are good models for structure, headers and footers, formatting and so on, particularly if you are printing your thesis two-sided.

   ***Allow much more time than you would imagine* for the final formatting processes and adjusting of the illustrations.** The amount of time needed takes most people by surprise.

Structure of a Thesis

Also read Chapter 1: *Structuring a Document, page 3*, for the following information:

1. The basic skeleton of section headings (page 3)
2. Choosing section headings: building an extended skeleton (page 5)
3. The *Outline* mode of Microsoft Word®: organising a document (page 11)
4. The importance of overview information: building a navigational pathway through your document (page 12)
5. Deliberate repetition of information in a document (page 14)

Elements Likely to Be Needed in a Thesis

The elements that a thesis should probably contain are those of the standard skeleton and several from the extended skeleton.

See the following:

- **The Basic Skeleton of Section Headings and Building an Extended Skeleton,** Chapter 1: *Structuring a Document*, page 3.
- **Other Possible Useful Sections for a Thesis,** this chapter, page 150.

Element	Purpose in a Thesis	Cross-Reference to Relevant Part of This Book
Title	To adequately describe the contents of your document in the fewest possible words.	See **Title**, Chapter 2: *The Core Chapter*, page 19
Main **Abstract** *or* **Summary**	To give the reader an overview of all of the key information in the thesis: objective, methods, results, conclusions, contributions to originality.	Chapter 3: *Abstract/ Summary/Executive Summary*, page 53
Acknowledgements	To thank your supervisors and the other people who have given you help in your research and in the preparation of your thesis.	See **Acknowledgements**, Chapter 2: *The Core Chapter*, page 23
Table of Contents	Gives the overall structure of the thesis. Lists the sections, chapters, headings and (*possibly*) subheadings, together with their corresponding page numbers.	See **Table of Contents**, Chapter 2: *The Core Chapter*, page 23
List of Illustrations **List of Figures** **List of Tables**	To give a listing – separate from the *Table of Contents* – of the numbers, titles and corresponding page numbers of all your figures and tables.	See **List of Illustrations**, Chapter 2: *The Core Chapter*, page 26
Glossary of Terms and Abbreviations *or* **List of Symbols**	To define the symbols, terms and abbreviations (including acronyms) that you use in the main text of the thesis.	See **Glossary of Terms and Abbreviations**, Chapter 2: *The Core Chapter*, page 27
Objectives	To give the main aims of the research.	
Introduction (*or* the introductory material under various headings)	• To clearly state the purpose of the study. • To allow readers to understand the background to the study, without needing to consult the literature themselves. • To describe the historical development of the topic. • To provide a context for the later discussion of the results.	See **Introduction,** Chapter 2: *The Core Chapter*, page 28 And **Other possible useful sections**, this chapter, page 150

Element	Purpose in a Thesis	Cross-Reference to Relevant Part of This Book
Literature Review (if the literature is not surveyed in the *Introduction*)	• To show that you have a good understanding of the historical development and current state of your topic. • To indicate the authors who have worked or are working in this area, and to describe their chief contributions. • To indicate correlations, contradictions and gaps in the knowledge, and to outline the approach you will take with respect to them.	See Chapter 4: *A Literature Review*, page 63
Other Sections Appropriate to the Topic		
Chapter Summaries	To give an informative (*not* descriptive) overview of the material in each chapter. For definitions of informative and descriptive, see Chapter 3, *An Abstract, a Summary, an Executive Summary,* page 53, and Chapter 6, *A Journal Paper,* page 88	See below, page 151
Overall Conclusions **Chapter Conclusions**	To give an overview of the conclusions drawn from (1) the whole work or (2) each chapter.	See **Conclusions**, Chapter 2: *The Core Chapter*, page 39
Recommendations (*if appropriate*)	To propose a series of recommendations for action.	See *Recommendations*, Chapter 2: *The Core Chapter*, page 40
Recommendations for Further Research (*if appropriate*)	To propose directions for further development of your work.	See *Suggestions for Future Work*, Chapter 2: *The Core Chapter*, page 41
List of References	A list of the works that you have cited in the text. Strict conventions govern this process.	See Chapter 15: *Referencing*, page 169
Appendices	For complex material that would interrupt the flow of the thesis if it were to be inserted into the main body, e.g. raw data, derivations, detailed illustrations of equipment, coding, specifications, product descriptions, charts and so on.	See *Appendices*, Chapter 2: *The Core Chapter*, page 42

Other possible useful sections for a thesis

(Note: All of the sections below can also be used as subsections of the *Introduction.*)

1. **Statement of the General Problem**
   A statement of the problem that the thesis work is designed to address
   or alternatively:
   Objectives or **Aim of the Study** (see **Objectives**, Chapter 2: *The Core Chapter*, page 30).
2. **State of Knowledge**
   A summary of the present state of knowledge in the area. Another name for a *Literature Review*, perhaps more appropriate to a major work such as a thesis. See Chapter 4: *A Literature Review*, page 63
3. **Contribution Summary**
   A summary of the areas of advancement or originality contained in the study.
4. **Scope of the Study**
   The areas that were and were not studied; the limitations of the study.
5. **Thesis Structure**
   A brief description of the various sections of the thesis and what they contain. This is likely to need descriptive statements rather than informative ones. See **Descriptive/Informative**, Chapter 3: *An Abstract, a Summary, an Executive Summary*, page 55
   Example
   Note the phrases used are those of descriptive rather than informative statements (*discusses, is stated, is described, deals with, gives a review of, chronologically surveys* and so on).

Thesis Structure

-
-
-

Chapter 3 discusses the integral boundary-layer methods used to calculate the viscous component of the study. The boundary-layer equations and other associated definitions are stated. The implementation of Thwaites' (1949) laminar integral boundary-layer method is described and validated against an experimental velocity distribution...

Chapter 4 deals with the interaction between the inviscid and viscous flow components. It gives an extensive review of available interaction schemes, the methods used to match the two flows and chronologically surveys over 30 viscous-inviscid interaction studies.

Chapter 5... etc.

6. A **Model Algorithm** as a map of the whole thesis.
7. **Each chapter should have the following:**
   a. **A chapter *Summary***
      - This gives an overview of the material in the chapter. See Chapter 3: *An Abstract, a Summary, an Executive Summary*, page 53
      - **Placed at the beginning or the end of the chapter?**
        The optimal position for the assessors is for the summaries to be placed at the beginning of the chapter because this helps them to assess the rest of the information (see **The Importance of Overview Information**, Chapter 1: *Structuring a Document*).
        Summaries have traditionally been placed at the end of chapters, but this is not the best position for the readers' understanding of the chapter.
      - ***Suggestion:*** also copy all of the chapter summaries to make a chapter of their own. This can provide a useful overview that is more detailed than the main *Summary*.

   b. **A set of *Conclusions***
      If appropriate to the material, a set of *Conclusions* at the end of a chapter reinforces the material in the assessors' minds and should form part of the intellectual pathway into the material in the next chapter.
   c. **At the very end of each chapter, probably immediately after the *Conclusions*, outline what the next chapter covers.**
      The assessors' understanding of the current chapter is helped by the knowledge of how this chapter will lead into the next one.

Checklists for the Sections of a Thesis

Use the various checklists for the relevant elements in the following:

Chapter 2: *The Core Chapter*
Chapter 6: *A Journal Paper*

14 A Conference Poster

This chapter covers:

- Attending a conference and presenting a poster: the basics
- Purpose of a conference poster
- Advantages and disadvantages of a poster
- What conference participants dislike in a poster
- Planning the poster
- Design of the layout
- Figures and tables
- Structure of the text
- Style of font
- Final production
- Checklist

For a conference, you can be asked to present either an oral presentation of your work (see Chapter 19: *A Seminar or Conference Presentation*, page 231) or a poster. A poster is a very common way of presenting work – particularly postgraduate work – at a conference.

There is a wealth of online information about poster templates and design. Because much of the design will involve your personal aesthetic judgment, this chapter covers only general principles and guidelines

Attending a Conference and Presenting a Poster: The Basics

For poster presentation at a conference, you may also be required to prepare a **conference abstract** and a **paper**. They are all different in their requirements:

1. **You will first be asked to submit an abstract of your work. The required length is usually between one and three pages.**
   See Chapter 3: *An Abstract, a Summary, an Executive Summary*, particularly **A Conference Abstract**, page 58.
2. **If you are accepted for the conference, you will then be invited to submit your paper.**
   This will be in the form of a standard scientific paper (see Chapter 6: *A Journal Paper*, page 83). You will be given a maximum page number and specific instructions about how to prepare it.

   The collected papers – usually from all of the participants and sometimes only from selected ones – will be published in the conference proceedings either in hardcopy or in digital format.

Writing for Science and Engineering.
DOI: http://dx.doi.org/10.1016/B978-0-08-098285-4.00002-9

3. **If you have been asked to present a poster**, you will be given the following information:
   - The poster dimensions. The usual format is A0 paper size: 118.9×84.1 cm (46.8×3.1 in).
   - The specific time and place in which you display it so that people can discuss your work with you. Most conferences will ask you display it for the whole of the conference and stand alongside for only a short specified time; the larger conferences may ask you to display it for only a short period of the total conference time.

Purpose of a Conference Poster

- To present the main points of your work as an enlarged graphic display.
- To give enough information to inform but to be simple, clear and creative.
- To present it so that it is visually pleasing and does not look too dense, ill-conceived or careless.

An effective conference poster is a blend of selected information and aesthetic design.

Advantages and Disadvantages of a Poster

Advantages

- Presenting a poster is a far less nerve-wracking experience than giving an oral presentation.
- Conference participants can choose to scan posters quickly or study them intensively.
- They are a visual medium and can be presented very attractively.
- You get personal contact with those interested in your work.

Disadvantages

- You do not have a captive audience, as in an oral presentation. You therefore have to attract people to your poster, which will be one of many in the same display space.
- Space is limited, so you have to select the information carefully.
- A poster can take more time and cost more to prepare than the visual aids needed for an oral presentation.
- You may feel somewhat deflated when many people drift past posters and take away very little impression of your work.

What Conference Participants Dislike in a Poster

Opinions that I've collected over 15 years from many hundreds of Ph.D. students have shown that they most dislike the following 12 aspects. This gives us some guidelines from which to work.

1. Too much text. This is by far the common and fervent complaint.
2. Text is too small.

3. No obvious logical flow of the information.
4. Too crowded.
5. Main points not clear.
6. No obvious 'takeaway message', i.e. your main conclusion(s).
7. Not enough illustrations.
8. Illustrations too small, too finely drawn.
9. Colour and/or background design interfering with the information.
10. Too many lines on graphs.
11. Photos or web material with not enough contrast or over-enlarged for their resolution.
12. Tables too large, too much information.

Planning the Poster

Planning what to include and how to lay it out are the two most crucial parts of producing a poster.

IMPORTANT: Don't just use your conference paper with perhaps with a few extra illustrations.

Some presenters adopt the quickest, least creative method and copy/paste their paper almost unaltered into the poster. The result is very poor: too much text, dense, squashed, overwhelming and discouraging to read.

Planning steps

Remember: What looks clear on your monitor can look very different when printed.

Step 1: Plan a tentative title.

(For information specific to a poster title, see *A Conference Poster Title*, Chapter 2, *The Core Chapter*, page 20.)

- This will need to be done long in advance of the conference, in some cases as much as 6 months.
- **The title should not only contain the key information but also draw the attention of the poster viewers.** For this reason, it can be shorter, punchier and possibly more querying or controversial than a title for a journal paper. However, questions as titles can not only be provocative, but they can also imply that your results are in question.

You also need to take account of its length. If it's too long, it will take up too much room because of the large font. In effective posters, the title font can be between 50 and 90 point type.

Step 2: Decide on your main points.

Define the 'take-home message', i.e. the main conclusions of your work. What do you *most* want people to remember about your work after viewing your poster?

Step 3: Make sure you know the size and orientation of your poster.

The organisers are likely to require A0; however, it could be different. The size will determine how much to limit the information; the orientation (portrait or landscape) will determine the layout.

Step 4: Decide on the illustrations to include, and then plan the poster around them so that they tell the story.

- Make the illustrations as self-explanatory as possible. Most viewers look more at the illustrations than the text.
- They should be simple and have proportionately chunkier lines and larger labelling than those in your paper.
- To avoid using too many words, use schematics and flow diagrams where possible.

Step 5: Choose a template.
You could use one from your presentation software or from the many online suggestions, or design your own. Some of the suggested standard templates are poorly designed and coloured.

Step 6: Select your information rigorously.
You will probably want to put more information into a poster than is realistic. You have to be extremely selective.

Step 7: Work out how you are going to show the logical structure of your material.

Step 8: Decide how to limit the amount of detail.
Too many presenters think that a poster packed with information gives the impression of productive research. It doesn't; instead, it's likely to obscure the central ideas. Remember that very few people will read every word of a densely packed poster.

Most viewers of a poster want something that looks clear and easy to absorb; however, there will always be a very few people who will argue for including lots of detailed material. The following are Arguments for limiting the information are:

- Participants' dislikes (see the list above, page 154).
- Viewers absorb the information much more readily if a few points are clearly stated and presented well, rather than given as a mass of detailed information.
- Level of lighting may be poor; background noise level may be high.
- Participants who are particularly interested in your work may have already read your conference paper and will want talk to you about the detail.
- Consider giving a handout sheet with more information instead of including it all on the poster.

Step 9: Plan for the text and illustrations to be easily read from a distance: the usual advice is from 2 m.
This means you must use a *large* font: 24 point for the main text is a good lower limit, 28 point is better. Be careful: this will probably look uncomfortably large on your monitor, so that you might opt for a smaller font that turns out to be too small.

Graphic artists tend to opt for a larger font size for effect, e.g. 30 point. Be careful; too large of a font size can make your poster look superficial. You need to tread a fine line between illegibility and superficiality.

Design of the Layout

A wealth of information is available online concerning layout design. It is difficult to give absolute advice about poster design. Some people have an instinctive feel about the aesthetics of design, and others do not. It's also a matter of personal preference.

You need to aim for a well-finished product, with an obvious logical flow and a good balance between text and illustrations. Suggestions:

1. **Place the title, author names, institutions and relevant logos all at the top of the poster.** All of these need to be clear; many participants will be interested in not only your specific results but also work from your institution or work group.

   The title itself should be in font large enough to be read from a greater distance than that of the main text of the poster: the usual advice is from 5 m away (say, 50–90 point).

   The authors' names should be in a smaller font than the title, perhaps two-thirds of the size. However, the author names are important and shouldn't be placed at the bottom of the poster or in small font. Don't forget that many of the participants will be searching out the names of specific people, e.g. your supervisor or principal investigator.

   The authors' affiliations (place of work) should be in smaller font than the authors' names, perhaps one-third of the title size.

   Make sure the poster heading containing this information does not take up too much space.
2. **Plan for the poster to be smoothly read from top to bottom and from left to right.** Don't make the viewers' eyes jump around.
3. **Boxes** are a good method of indicating the logical flow of information to the viewer.
   a. **Group the information into boxes, together with the relevant illustrations** (Figure 14.1).
   b. **Inside the boxes, use numbered, informative headings with the following characteristics:**
      - Numbered because this is an effective way of showing the flow of information.
      - As far as possible, make each box heading give the main conclusion of the work in that box. The boxes serve as a form of summary information and help the viewer's understanding.

   Example:

   Stimulation alters protein distribution in functional vesicle clusters

   is more useful than…

   Protein distribution in functional vesicle clusters

5. **If you are not using boxes, heading bars** can show the width of information under one heading (Figure 14.2).
6. **Make sure that your text is not dense, as in Figure 14.2.** Very few people will bother to read it all. Aim for the minimal amount of text together with excellent self-explanatory illustrations.

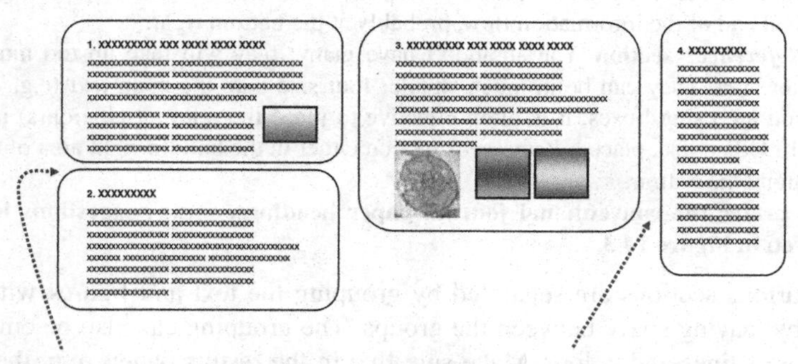

Figure 14.1 Use of boxes to group information.

Methods | **Results**

Figure 14.2 (1) To show use of heading bars; (2) the visually poor effect of making text too dense.

7. **Arrows: be very careful.** The flow is usually obvious to the author, but too many or multi-directional arrows are confusing and often appear less than logical to the viewer.
8. **Suggested headings.** Your choice of section headings is far less constrained than when writing a paper. There is no need to use the rigid, classic *TAIMRAD* headings (*Title, Abstract, Introduction, Methods, Results and Discussion*). However, you need to follow that logical development of the information. Suggestions:
   - **Is an *Abstract* needed?** Some people think a poster needs an *Abstract*; some think that because a poster is condensed information, it isn't necessary. Check with your supervisor.
   - **Make sure that the reader is in no doubt about the objectives of the work.** This will increase understanding of your work. Either include this in the *Introduction* box, possibly as a separate side heading, or give it a small box to itself.
   - **Is a section describing your future work needed?** Some supervisors may be unhappy about indicating how your work is going to be developed: there are instances of competitive groups taking the idea and publishing before you.
   - **A *Conclusions* section is absolutely necessary**. It should be made up of very brief, listed conclusions from your work. This is the information that sums up your poster and that the viewer will use to appraise your work – the take-home messages. Place it at the logical end of the information flow, probably at the bottom right.
   - **A *References* section**. You shouldn't have many; they will take up too much of the poster area. They can be in much smaller font size than the main text (e.g. 14 point). If you are using boxes, it is often effective to place the relevant citation(s) in the box itself. Otherwise, place a *References* section either in the bottom-right area of the poster or along the bottom.
9. **If you prefer the conventional journal paper headings, some suggestions for layout are given in Figure 14.3.**

The various sections are separated by grouping the text and figures within each one and by leaving space between the groups. The grouping can also be emphasised by the use of lines and colour. Make sure that in the results panels, e.g. the reading direction is obvious to the viewer.

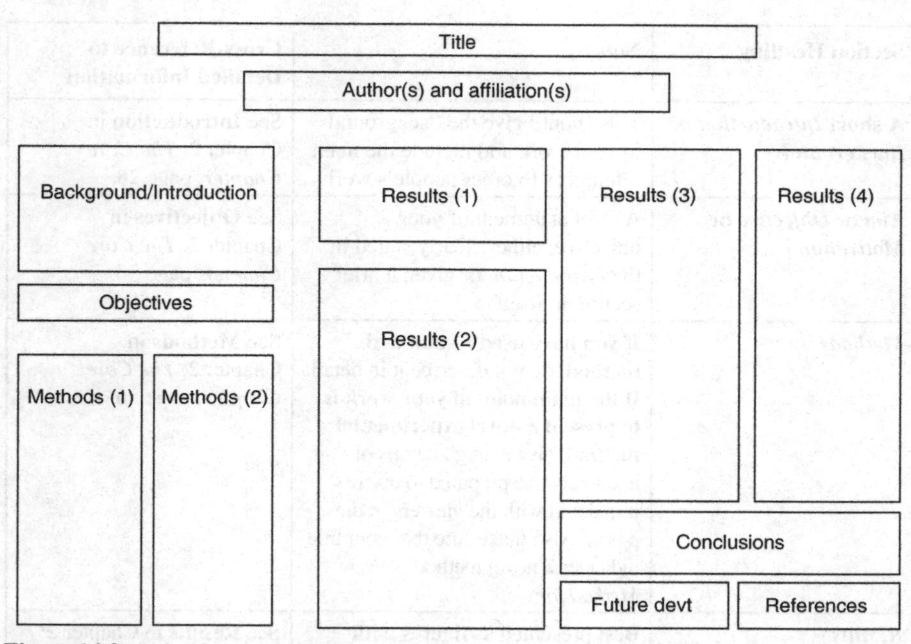

Figure 14.3 A simple, conventional poster layout where the flow of information is obvious.

If you do want to use some of the *TAIMRAD* headings, the following table gives some information about how to use them in a poster and the relevant cross-references to more detailed information in the book.

Section Heading	**Notes**	**Cross-Reference to Detailed Information**
Title **Author(s)** **Place of work (usually called affiliation)**		See **Title**, Chapter 2: *The Core Chapter*, page 17. In particular, **A Conference Poster Title**, page 20. See **Authorship and Affiliation**, Chapter 6: *A Journal Paper*, page 87.
A very short *Abstract*, if required. Many people think that an ***Abstract*** is not necessary in a poster.	The abstract should be **very short**: Make sure it contains the bare outline of the **methods** you used, together with your **results** and **conclusions**. Place it in the logical position where people will expect it: at the top, immediately under the title/authors/affiliation, probably to the left.	See Chapter 3: *An Abstract, a Summary, an Executive Summary*, page 53.

Section Heading	Notes	Cross-Reference to Detailed Information
A short *Introduction or Background*	This should give the background to your work and include the main references to other people's work.	See **Introduction** in Chapter 2: *The Core Chapter*, page 28.
Aim or *Objective* or *Motivation*	A brief statement of your objective, either clearly stated in the *Introduction* or given a brief section to itself.	See **Objectives** in Chapter 2: *The Core Chapter*, page 30.
Methods	**If you have used a standard method**, do not describe it in detail. **If the main point of your work is to present a novel experimental method**, give enough details of it, and also be prepared to discuss it in detail with the viewers of the poster. Also make sure that your title indicates a novel method (*A New Method for …*).	See **Methods** in Chapter 2: *The Core Chapter*, page 36.
Results	Best presented as figures with a small amount of linking text. Avoid tables if possible, unless they are absolutely necessary and very simple.	See **Results** in Chapter 2: *The Core Chapter*, page 37. See **Illustrations** in Chapter 2: *The Core Chapter*, page 44.
You may want a short **Discussion** section. It may not be necessary.		See **Discussion** in Chapter 2: *The Core Chapter*, page 38.
A brief ***Conclusions*** section	It should be in the form of a short list, concisely worded, starting with your major conclusion. Position it at the logical end of the information flow, probably at the bottom right or centre.	See **Conclusions**, Chapter 2: *The Core Chapter*, page 39.
Future Development	It may be particularly relevant at a conference to show how you are planning to extend this work. However, some supervisors are reluctant for this to be included because of possible pre-emption by other groups.	See **Suggestions for Future Development**, Chapter 2: *The Core Chapter*, page 41.
A very short ***List of References***, if your poster cites other people's work.	Place in small font either in the relevant boxes or at the bottom of the poster.	See Chapter 15: *Referencing: Text Citations and the List of References*, page 169, for the conventions.

Figures and Tables

Viewers look first at illustrations

Very often, the illustrations are the only part of a poster that people really study. They should be well-presented, clear and readily understood as far as possible without reference to the text. They should be placed into the relevant boxes (see *Design of the Layout*, page 156 above).

If you are not using boxes, make sure that the illustrations are not scattered around and placed far away from the relevant text citations.

See ***Illustrations***, Chapter 2: *The Core Chapter*, page 44, and ***Figures for a Journal Paper: General Guidelines***, Chapter 6: *A Journal Paper*, page 102.

Make your illustrations look outstanding

There will be other students in your institution with experience in using software to produce superb illustrations. Don't be afraid to ask them for advice.

Guidelines for illustrations

1. **Plan the poster around the figures: use them to tell the story**.
   Self-explanatory and well-placed figures need very little linking text to be able to convey the storyline.
2. **Make each figure self-explanatory**.
   If you are using boxes Figures in the relevant box are more convenient for the viewer than having to move from figure to text and back again to be able to understand the argument.
   If you are not using boxes: Figures should be numbered and positioned as close as possible to where they are mentioned in the text. If the reference in the text to the illustration is made prominent (e.g. by being in bold, uppercase or italics), it is easier for the viewer to cross refer from the text to the illustration or vice versa.
3. **Schematic diagrams of equipment and procedures** are a good way of avoiding too many words. They are particularly effective in the *Introduction* or for describing methods.
4. **Graphs**
   - Make the points, axes and lines clearly visible from 2 m away.
   - Don't use enlarged copies of your paper's graphs. They will appear too thin and spidery.
   - Don't have too many lines on one graph.
   - Avoid red and green lines. Colour-blind people will not be able to distinguish between them.
5. **Tables**
   Avoid tables if possible. Any tables should contain far less information than those in written documents. They are usually not effective on posters, unless they are extremely simple. If possible, present the information in other ways.
6. **Photographs**
   Make sure that they have good contrast and are not enlarged beyond their capabilities. Unclear or pixelated illustrations look very unprofessional.

Style as in Written Text	Short, Simple, Separated Text Suitable for a Poster
Sandwich composites are used on aircraft because of excellent stiffness-to-weight ratios. However, they have low damage tolerance and are frequently impacted in normal operation because of their locations in the aircraft. To date, virtually no information has been available on the effects of impact damage.	**BACKGROUND** Sandwich composites on aircraft: • have excellent stiffness-to-weight ratios • have low damage tolerance • are frequently impacted because of their location. To date, virtually no information has been available on the effects of impact damage.

Figure 14.4 Example of text style as in a conference paper and the same material presented for a poster. Text can be understood much more quickly when it has been separated: here, empty space and a variable left margin have been created by indenting and listing within the text.

Structure of the text

Viewers often scan the text quite rapidly. Therefore:

- **Don't use longish paragraphs as in your paper.**
  Use short, simple and separated statements (Figure 14.4).
- **Use informative headings as much as possible, i.e. headings that state the conclusion of the information below each.**
- **Differentiate the headings from the text by using**:
  UPPERCASE *or* **Boldface** *or* SMALL CAPITALS or A DIFFERENT FONT

Style of font

- **Use a simple font.**
  Elaborate fonts are difficult to read and can look unprofessional.
- **Don't use too many fonts.**
  The usual guideline is no more than two, but use them discreetly; e.g. one font for the text and one for the headings.
- **Serif or sans-serif font?**
  A serif font is one that has fine ticks on the letters, e.g. Times Roman or Palatino. Commonly used in documents, serif fonts are usually regarded as too elaborate to be used in a poster.

A sans-serif font is simple, clean and more easily read at a distance and looks more professional for a poster. There are obvious differences in letter shape and the amount of space they use up; e.g. Calibri is significantly smaller and less dense than Arial or Verdana (see Figure 14.5). Be careful of *Comic Sans* which is a sans-serif font that students love, and many senior academics think looks frivolous.

- **Boldfacing, UPPERCASE and *italics*.**
  Use **boldfacing** only for special emphasis, such as the title and headings. Solid blocks of text in boldfacing can look harsh and be difficult to read.
  Use **UPPERCASE** only for headings. Avoid blocks of text in UPPERCASE or **italics**; they are known to be difficult to read and discouraging.

Using colour

Use colour inventively but carefully:

- **Unify the poster elements** with an intelligent use of colour.
- **Make your poster eye-catching.** We are visual people, so colour attracts our attention. But don't make it too raucous; you'll want it to be seen as professional work.
- **Don't use too many different colours.** Vivid, clashing schemes might work; toning colours that complement each other can produce a very elegant result.
- **A good way of introducing colour without it interfering with the information** is to use white-filled boxes for the information and plain or shaded colour for the background.
- **Shaded backgrounds** Beware of shaded backgrounds, if material is superimposed on them. The variations in colour can badly affect the clarity of the information. **Be aware of red superimposed on a bluish background.** Even though it may look clear on your monitor the red becomes difficult to distinguish on the poster.
- **Photos can be used as background**, but even when made transparent, they can badly compromise the information because of the colour and pattern variations.

Overall rule: Choose the background and colour combinations so that the text and illustrations stand out clearly.

Using presentation software

A presentation software package such as Microsoft PowerPoint® can produce superb poster material. However, consider the following:

- Beware of some of the presentation software standard templates: they can be fussy and ill-coloured. Choose carefully and customise the colour, or design your own scheme.
- The standard templates are easily recognisable. Viewers can get the impression of a hasty job.
- Don't be tempted to use the Clip Art images. They are not appropriate in a formal science presentation.

Final production

The background is often provided at the conference.

Most conferences provide free-standing surfaces made up of cork board or fibreboard. You'll be provided with the means of attaching the poster.

For transporting your poster, use a poster tube.

PROPERTIES Sandwich composites have excellent stiffness-to-weight ratios.	Arial, Helvetica
PROPERTIES Sandwich composites have excellent stiffness-to-weight ratios.	Verdana
PROPERTIES Sandwich composites have excellent stiffness-to-weight ratios.	Calibri
PROPERTIES Sandwich composites have excellent stiffness-to-weight ratios.	Comic Sans Beware: many senior academics think it looks frivolous.
PROPERTIES Sandwich composites have excellent stiffness-to-weight ratios.	Univers condensed
PROPERTIES Sandwich composites have excellent stiffness-to-weight ratios.	Times New Roman. Serif font: usually regarded as not suitable for posters—difficult to read.

Figure 14.5 The difference in letter shape and line spacing of various sans-serif fonts suitable for posters, and a serif font for comparison. They are shown in boldface and all are the same font size.

Common mistakes

1. Too much text: this is very common (*as in manuscript*)
2. Too much information is given.
3. Main points are not clear.
4. No clear conclusions are provided.
5. The flow of information is not obvious.
6. If using arrows, the direction of flow is not clear to the reader. Avoid arrows if possible.
7. Too much information is crammed in to attempt to cover too many points.
8. Too much detailed information is provided.
9. The background and/or colour interfere with the information.
10. The font is too small and cannot be read from 2 m.
11. Font style is inappropriate for a poster.
12. Tables contain too much information.
13. Illustrations are too finely drawn and therefore difficult to see.
14. A lack of planning is obvious.
15. Photographs and online illustrations are enlarged beyond their capabilities.
16. Material is too dense causing a lack of empty space.

Checklist for a conference poster

- ☐ Have you planned the poster around the illustrations?
- ☐ Have you avoided trying to present too much information?
- ☐ Does it look as though the information is easily extractable by a viewer?
- ☐ Is the flow of the story self-evident to a viewer?
- ☐ Does the poster look crammed with too much information?
- ☐ Is there only a relatively small amount of text?
- ☐ Is the font size easily readable from 2 m away?
- ☐ Is the text in a simple sans-serif font (*Arial, Univers, Avant Garde*)?
- ☐ Are you sure you want to choose a serif font (*Times Roman, Palatino*), even though most people don't like it in a poster?
- ☐ Is colour used inventively and intelligently?
- ☐ Does the background design and/or the colour interfere with the information?
- ☐ Illustrations:
  - ☐ Are they self-contained, with self-explanatory titles and captions?
  - ☐ Can they be understood in overall terms without needing to refer to the text?
  - ☐ Are the lines thick enough to be seen and the labelling clear?
  - ☐ Have you avoided squeezing the illustrations into spaces left between the text items?
  - ☐ Have you avoided tables if possible? If a table is needed, is it simple?

Section 3

Referencing, Editorial Conventions, Revising, Proofreading

15 Referencing: Text Citations and the List of References

This chapter covers:

- **How to cite your sources (called *references*) in the text.** The types of sources include the following:
  - Works on paper (books, journals, etc.)
  - Electronic sources (e.g. Internet material, electronic databases)
  - Other types (e.g. video and audio material)
- **How to present the section called *References* or *List of References*** and/or a *Bibliography*.

Note: This chapter assumes no prior knowledge of this area.

You may feel inadequately prepared for referencing. It is one of the most convention-ridden areas of scientific and technological writing. Many course notes give only bare outlines of how to do it.

This chapter assumes you have no previous knowledge of referencing. The aim is to give you all of the competency needed in this area.

General Guidelines

What is referencing?

Referencing is a system of referring to other people's work in a document that you're writing. Many authors of professional reports need to do this.

Why you have to be exact when you document?

It is essential that the sources of all your factual material are acknowledged in an accepted format. It's very easy to get the details wrong.

Most staff assessors are extremely meticulous about the way sources are referenced in assignments and will check your work very thoroughly.

Writing for Science and Engineering.
DOI: http://dx.doi.org/10.1016/B978-0-08-098285-4.00006-6

Why use referencing?

- To acknowledge other people's work or ideas in relation to your own.
- To enable readers of your document to find your source material.
- To avoid plagiarism or literary theft.

Failure to acknowledge sources is plagiarism and is a form of stealing. People who do not fully acknowledge their sources, or copy text word for word from them, are implicitly claiming that the work is their own. Universities have strict disciplinary procedures regarding plagiarism. You risk failing your assignment, exclusion from the course and, sometimes, suspension from university (see page 188, *this chapter: How to Avoid Plagiarism*).

When will you need to use referencing?

You need to use references in the following situations:

1. **When you write documents – such as reports – that refer to factual material taken from other sources.** This is the commonest form of documentation in a science or engineering assignment. It is the form used almost exclusively when you write papers for journals and is therefore the form that is monitored the most critically by staff.The sources may include the following:
   a. **Material on paper** (many of which are also online) such as:
      - Papers in professional journals and conferences
      - Books or book chapters
      - Theses
      - Lecture or laboratory documents
      - Magazine articles
      - Newspaper articles
      - An organisation's publicity material
      - Engineering standards and specifications
      - Government documents, such as Acts of Parliament and reports of committees
      - Others
   b. **Electronic sources** such as:
      - Internet (web) pages
      - Databases
   c. **Visual and audio material** such as:
      - Movie clips, DVDs
      - CDs
2. **When you need to quote word-for-word from another work** (see page 186 *Using Direct Quotations*).
3. **Differences in Documentation Between Arts-Related Subjects and the Sciences**

In arts-related subjects, you may have written essays that required a footnoting system using *ibid.* and *op. cit.* to cross-refer to previously cited sources. This system is not used in the referencing process for scientific and technological literature. Neither are footnotes, except in very rare cases.

Why must documentation be so thorough?

- **To avoid plagiarism.** If you include factual material in your writing that has been taken from sources such as books, the Web, magazines, etc., and you do not say where you got it, your reader/assessor is justified in the following actions:
  - Not trusting the material because there is no verifiable source of it.
  - Accusing you of plagiarism, or literary theft. People who do not fully acknowledge their sources are copying the work of others and implicitly claiming that the work is their own. Students risk failing their assignment, exclusion from their course and, sometimes, suspension from university (see page 188, *this chapter: How to avoid plagiarism*).
- **To put your work in context**. All scientific and technological work must be put in the context of other work in the field. Your reader needs to know that you are familiar with the literature in your area and that you can assess your work in relation to it.
- **To enable other people to follow up the reference** if they wish. This means it must be cited accurately and in detail.

What is a bibliography?

A Bibliography is a list of all the sources you have consulted while writing your document, only some of which are cited in the text.

Most institutions in science and engineering require a *References* section and very rarely ask for a Bibliography. A *References* section shows that you are familiar with the literature and can cite it appropriately in your own work. **It is essential to find out what your institution or journal requires.**

The Basics of Referencing

1. **There are two linked elements to referencing a technical document:**
   - **The sources that you used in preparing your document** (websites, books, articles, etc.) **are cited at the appropriate places in the text.**
   - **All the sources are then listed at the end of your document in a section called *List of References*** (can also be called *References*).
2. **There are two basic systems of referencing technical documents used in science and engineering:**
   - The author/date (Harvard or APA) system
   - The numbering system

   Note: Librarians will refer to these systems and their variations by various names. However, this book uses for the two broad systems the names *Author/date* and *Numbering* because the format of each is easily understood from the names.
4. **Referencing is one of the most convention-ridden areas of scientific and technical documentation**. Many assessors expect the conventions to be observed in the minutest detail.

The two main systems of referencing

There are two main systems commonly used in technical documentation for cross-referencing citations in the text with the full reference in the *List of References*. The two systems are described in overview in Table 15.1.

Most institutions prefer one or the other system. The following are two very important points concerning these systems:

1. **Always use the system that the journal or your journal or institution requires.**
2. **Never use a mixture of the two systems in any one assignment.** It is essential that you use only *one* system of referencing in a document – *either* author/date *or* numbering.

Table 15.1 Overview of the Two Referencing Systems

Author/Date System	**Numbering System**
In the text of the document • Surname of the author and the date of publication placed in parentheses. *For example*: (Brown, 2012).	**In the text of the document** • Each citation in the text is given a unique number, either in square brackets, e.g. [5], or superscripted, e.g. ^{5}. Each is numbered in the order in which it appears in the text. • If you need to cite a reference more than once in the text, the number of its first appearance (its unique number) is used each time you cite it.
List of References • Listed in alphabetical order of the surnames of the authors.	***List of References*** • Not listed alphabetically. It is a list numbered from 1 to *n*, the number of each listing corresponding to the unique number that each source was assigned in the text.

Choosing between referencing systems

Table 15.2 provides the pros and cons of the two systems.

Table 15.2 Advantages and Disadvantages of the Two Referencing Systems

Author/Date System	**Numbering System**
Advantages • Allows the source to be recognised by author and date in context within the text of the report (*Note: This is seen as a considerable advantage by people familiar with the literature.*) • Provides an alphabetical list at the end of the document. • Inserting an extra reference into the text is easy.	**Advantages** • The text of the document is not interrupted by wordy citations. • Only a number needs to be repeated: prevents repetition in the text of the same wordy citations.

Table 15.2 (Continued)	
Author/Date System	**Numbering System**
Disadvantages • Can create disruption to the text when there are many citations in one place.	**Disadvantages** • While reading the text, readers familiar with the literature cannot recognise the work that you are citing. They have to turn to the *List of References* to match a numerical reference to its source. • It can be difficult to add another citation and renumber all successive ones. But this can be overcome by using endnoting software. • The numbers give no information about the work, and it is easy to forget to use the earlier number when you need to refer to it again later in your report. Again, endnoting software will overcome this.

Citing references in the text

Author-date system

Overview:

- The sources cited in the text are in the form of (*Author, date*) e.g. **...as has been previously noted (Brown, 2012)**

or

- If the author's name occurs in the text, the date follows it in brackets e.g. **...as previously noted by Brown (2012)**

Author's surname and date are placed in brackets.	Tewari (2012) showed that sulphur deficiency caused paling of the youngest emerging leaves.
Author's surname is cited in the text.	McGill (2012) has proposed that all six unified theories use the same three rules or assertions to describe a stochastic geometry of biodiversity.
References are precisely placed.	This runoff has also introduced heavy metals (Davies, 2012), pesticides (Schultz, 1998), pathogens (Cox, 2006), sediments (Horb, 2001), and rubbish (Williams, 2009).
The source is by two authors.	Martin and Zubek (1993) compiled a comprehensive list of dust activity on Mars, from 1983 to 1990. Regular dispersion patterns will result if communities comprise groups of organisms that use different components of the physical space (Henderson and Magurran, 2012).

The source is by more than two authors. Cite the surname of the first author and add "et al." (italicised in some house styles).	Zuidema et al. (2012) have shown that recruitment subsidies can be crucial for maintaining subpopulations of tropical tree species.
Several sources are cited within one set of brackets. Depending on house style: separate them by semicolons, and cite them in order of *either* (1) publication date *or* (2) by alphabetical order of the author.	The locomotion activity of a given species may be a source of considerable error in estimating energy budgets (Boisclair and Sirois, 1993; Facey and Grossman, 1990; Hansen et al., 1993; Lucas et al., 1993; Ney, 1993; Ware, 1975).
Two or more papers are written in different years by the same author.	If the interfacial shear stress is assumed to be constant, the recovery length is related to the maximum shear stress in the fibre (Curtin, 1991, 1993).
The author has written several papers in one year. Distinguish between them by adding a lowercase letter to each paper. These letters must be added to the listing's date in the *List of References*.	Previous analysis of the *Clock* gene in mice (King et al., 1997a, b) has shown that *Clock* is expressed in a manner consistent with its role in circadian organisation. In mice (King et al., 1997b) the CLOCK locus lies distal to...
There is a large body of work, but you are citing only a few representative examples. Use e.g. within the brackets.	Martian dust storms, also called Martian yellow storms or Martian yellow clouds, have been observed for a long time (e.g. Antoniadi, 1930; Peters and Petrenko, 2009).
Reference a large body of information contained in a review paper.	Zebra fish generate large numbers of transparent embryos that develop synchronously to a free-swimming hatchling in a period of three days (for review, see Driever et al., 1994).
Cite a major source a number of times (e.g. a textbook).	... (Clarkson, 2002, p. 51)
Although it is not part of the conventions, it is useful to the reader if the individual page numbers are cited with each text reference.	*or* Clarkson (2002, p. 52) stated that...
You have been unable to obtain the original reference but have seen it cited in another paper.	*Smith (1928) as cited by Brown (2001)*... would be appropriate if you learnt about Smith's paper through Brown's, but have been unable to read Smith's.
It is acceptable to cite the secondary source provided the primary source is included.	In the *List of References*, give full citation details of both.
Different authors with the same surname have published in the same year.	It has been shown by Smith, C.W. (2008)... However, Smith, J.G., (2008) reported that...

You only know the publication date of the source approximately. Use a small c before the date.	All the branches of a tree at any degree of height, if put together, are equal to the cross-section of its trunk (Leonardo da Vinci, c. 1497).
The author is not stated in the source, including electronic sources with no cited author (see the section on *Electronic Sources*, page 181). Use the first few words of the title and the date if known. For example, where the citation is: • Avon River Intake Feasibility Report (2012). Roberts Consultants Ltd., Contract TKA 2012/101. Prepared for Middletown Central Electricity Generation. • *The Virtual Trebuchet*. Retrieved May 8, 2012 from http://heim.ifi.uio.no/~oddharry/blide/vtreb.html • Hazelnuts worldwide (undated). Hazelnuts International S.A.	… (Avon River Intake) … (*The Virtual Trebuchet*) … (Hazelnuts worldwide)
The material is undated. Use *(undated)*.	Rivers and streams in the Hutchison Ranges (undated). Middletown City Council.

Copying or adapting illustrations (when using the author/date system)

At the end of the title for the figure, insert the appropriate phrase as shown below:

You have used an exact copy of an illustration from someone else's work.	**Figure 4** Schematic of the production line at FlatPack Furniture Ltd. (Reproduced from Pinkerton, 2012).
You have redrawn an illustration from someone else's work.	**Figure 1.2** Schematic of the production line at FlatPack Furniture Ltd. (Redrawn from Pinkerton, 2012).
You have adapted someone else's data or figure, and incorporated it into a figure or table of your own.	**Figure 3.** Schematic of the production line at FlatPack Furniture Ltd. (Adapted from Pinkerton, 2012).

Personal communications

If someone has told you or written a note to you about an aspect of your work, it is quoted as *pers. comm.* Cite the initials and the surname. Note: Personal communications are not usually included in the *List of References* section.

The sample was maintained at 25°C and pH 5.0 (D.J. Wilson, pers. comm.).

However, if you have a number of them and to give them authenticity, it may be appropriate to have a separate section for them. Place it after the *List of References*, headed *List of Personal Communications*. This should list in alphabetical order the surnames, initials and places of work of the people cited. You may also want to include the means of communication and its date.

Example:

List of Personal Communications

1. Broom, J.D. Department of Chemistry, University of Middletown. By email, 14/4/2012.
2. Simmonds, W.G., Department of Sport Science, University of Technology, Middletown. In discussion, 23/4/2012.
3. etc.

Numbering system

Overview	
• Each source cited in the text is given a unique number, in the order in which each is cited. • If you need to cite a reference more than once in the text, the number of its first appearance – its unique number – is used each time you cite it.	The wind velocity and behaviour of a geographical region is a function of altitude, season and hour of measurement [1]. Mylona [2] has analysed changes in sulphur dioxide and sulphate concentrations in air over a seven-year period. *or* The wind velocity and behaviour of a geographical region is a function of altitude, season and hour of measurement[1]. Mylona[2] has analysed changes in sulphur dioxide and sulphate concentrations in air over a seven-year period.

How to Compile the *List of References* Section

The *List of References* section is made up of a list of the papers, books, articles etc. that you have cited in the text of your work. It is placed at the end of your document, before any appendices (see Chapter 2: *The Core Chapter*).

Author/date system: List the references in alphabetical order of the surname of the author or first author if there is more than one. Sample text and *List of References*: page 182.

Numbering system: List the references in numerical order according to the unique number each source has been assigned in the text. Sample text and *List of References*: page 184.

Points to note:

- **Each reference is listed only once**. This is important.
- **There are minor variations** in the way the lists are cited for different house styles, for example, in the position of the date, the use of italics, quote marks and so on.

It is important to find out exactly the form that your institution requires and to stick to it rigidly.

- **Be sure that every full-stop or comma is in the right place, and all other aspects of the formatting are correct.** Formatting of references is riddled with convention, and academic staff members often check this area very thoroughly.
- **There are standard abbreviations for the journals.**

Don't make them up. The following two useful sites provide the information you need:

1. Science and Engineering Journal Papers: http://www.library.ubc.ca/scieng/coden.html
2. Journal Title Abbreviations: http://library.caltech.edu/reference/abbreviations/

Examples of how to list the various types of sources

A generalised scheme is shown here. But be aware that there can be minor variations in order and formatting of the individual items; it depends on the house style of the institution or journal.

Books

- **Surname and initials** of the author(s) or editor(s) (surname first, followed by the initials). If editor, place Ed. after the initials.
- The **year** of publication.
- **Title** of the book underlined or in italics, and with the 'main' words (everything except articles, prepositions and conjunctions) capitalised. For the conventions, see *Written Style for Headings*, Chapter 16: *Conventions Used in Scientific and Technical Writing*, page 196.
- If there is a **subtitle**, it is separated from the main title by a colon (:)
- **Title of series**, if applicable.
- **Volume number** or number of volumes, if applicable.
- **Edition**, if other than the first.
- **Publisher.**
- **Place of publication** (city or town).
- **Page numbers** of the material cited (if applicable).

Book with one author	Cassedy, E.S. (2006) *Prospects for Sustainable Energy*. Cambridge University Press, Cambridge.
Book with more than one author	Mitchell, W.J., Borroni-Bird, C.E., and Burns, L.D. (2012) *Reinventing the automobile: personal urban mobility for the 21st century*. MIT Press, Cambridge, Mass.
Book with one or more editors	Draelos, Z.D. (Ed) (2012) *Cosmetic dermatology: products and procedures*. Wiley-Blackwell, Chichester.
One volume of a multi-volume work	Erdélyi, A. Ed (1955) *Higher Transcendental Functions*. Vol. 3. McGraw-Hill, New York.
Second or later edition of the book	Barrett, C.S. and Massalski, T.B. (1980) *Structure of Metals: Crystallographic Methods, Principles and Data*. Third edition. Pergamon Press, Oxford.
A chapter or article in an edited book	Kenzel, W. and Opper, M. (1991) 'Dynamics of learning'. In: *Models of Neural Networks*. Eds: Domany, E., van Hemmen, J.L. and Schukten, K. Springer-Verlag, Berlin, pages 99–120.

• The chapter title is enclosed in quotation marks. • The name of the book is preceded by In:. • Starting and ending page numbers of the chapter are given.

Journal papers

- **Surname and initials** of the author(s) (surname first, followed by the initials).
- The **year** of publication in parentheses ( ).
- **Title** of the paper.
- The **name of the journal**, *in italics* in its correctly abbreviated form (see Points to note, page 177, above).
- The **volume number** of the journal, usually in **boldface** (with the issue number, if there is one, in brackets; *see examples 2–3 below*).
- The **numbers of the pages** on which the paper begins and ends. ***Note***: The actual page from which your information is taken is not cited.

Single author	Amos, A.J. (1958) Some lower Carboniferous brachiopods from the Volcan Formation, San Juan, Argentina. *J. Paleontol.*, **32**, 838–845.
Two authors	Franceschini, S. and Tsai, C.W. (2012) Assessment of uncertainty sources in water quality modelling in the Niagara River. *Adv. Water Res*, **33** (4), 493–503.
Multiple authors	Walsh, C. J., Roy, A.H., Feminella, J.W., Cottingham, P.D., Groffman, P.M., Morgan, R.P. (2005) The urban stream syndrome: current knowledge and the search for a cure. *J. N. Am. Benthol. Soc.*, **24** (3), 706–723
Paper in the proceedings of a conference As for a journal paper but in addition, state the **number** of the conference, its **title theme**, the **place it was held** and the **date**.	Bhattacharya, B., Egyd, P., and Toussaint, G.T. (1991) Computing the wingspan of a butterfly. Proc. Third Canadian Conference in Computational Geometry (Vancouver), Aug 6-10, 88–91.
Paper in language other than English, not translated	Schmutz Schaller, P. (2000) Platonische Koerper, Kugelpackungen und hyperbolische Geometrie, *Math. Semesterber.* **47**, 75–87 (in German).
Put (in *language*) at end of the citation. The title may remain in the original language, or be translated into English.	Gorb S.N. (1989) Functional morphology of the arrester-system in Odonata. *Vestn. Zool.* **89**, 62–67 (in Russian).

Other types of sources

Note: If the author is not stated, do the following:

Describe the source as fully as possible, in the style of the following relevant examples. The order of the items cited is given here:

1. The title of the document
2. Date (when possible)
3. The organisation/institution that produced the document
4. Any identifying number, such as designation code, or contract number

For citation in the text: Use an abbreviated form of the title (see above, page 177).

Thesis	Johnson, C.E. (2001) A Study of Residual Stresses in Titanium Metal Matrix Composites. PhD Thesis, University of Middletown.
Student project	Young, E.A. (2002) Mathematical modelling of land-mine detection. Engineering Science project, School of Engineering, The University of Middletown.
Lecture material	If the writer's name is stated: Carter, R. (2002) Robotics. Lecture handout, *Engineering and Society*, The University of Middletown. If the writer is unknown: Wetlands (2002). Lecture handout, *Conservation Ecology*, The University of Middletown.
Laboratory manual	Strain measurement (2002). Year Two Mechanical Engineering Laboratory Manual, The University of Middletown, 46–49.
Newspaper article	*Author is known*: Nicholson-Lord, D. (1995) Does work make you stupid? Independent on Sunday, 29 January, p 21. *Author is unknown:* Could alcohol be good for your liver? The Week, 13 November 1999.
Magazine article	*Author is known:* Crystal, D. (1999) The death of language. Prospect, November 1999, 12–14. *Author is unknown:* How clean is your water? (2002). Water News, Number 15, 19–21.
Technical report	Hilley, M.E. Ed. (1971) Residual Stress Measurement by X-Ray Diffraction. SAE Information Report J784a, Society of Automotive Engineers, New York.
Microfiche	Buckley, D.H. (1985) Tribological Properties of Structural Ceramics. NASA, Washington DC. Microfiche.

Government and legal documents • The first element of information is the government department, committee or body. The last two may also be referenced by the name of the chairperson. • Include the complete title.	CORINAIR Working Group on Emission Factors for Calculating 1990 Emissions from Road Traffic, 1 (1993). Commission of the European Committees (Office for Official Publications, Luxembourg).
Section of an Act of Parliament	Risk assessment and notification requirements (1990) Environment Protection Act 1990 (c. 43), Part VI – Genetically Modified Organisms, Section 108. Act of Parliament, United Kingdom. Her Majesty's Stationery Office, London.
Report by a professional body	Recycling Household Waste – The Way Ahead (1991). Association of Municipal Engineers, The Institution of Civil Engineers, London.
Engineering codes	Building Code Requirements for Reinforced Concrete and Commentary (1989). ACI Committee 318, American Concrete Institute, Detroit.
Standard specification	Standard Specification for Urea-Formaldehyde Molding Compounds (1994). Designation D705-94. American Society for the Testing of Materials, Annual Book of ASTM Standards 1999. **08.01** *Plastics (I)*, 92–93.
Standard test method	Standard Test Methods for Thermoplastic Insulations and Jackets for Wire and Cable (1996). Designation D2633-96. American Society for the Testing of Materials, Annual Book of ASTM Standards, 1998, **10.02** *Electrical Insulation (II)*, 25–38.
Standard practice	Standard Practice for Algal Growth Potential Testing with *Selenastrum capricornutum* (1993). Designation D-3978-80 (Reapproved 1993). American Society for the Testing of Materials, Annual Book of ASTM Standards 1997, **11.05**, *Biological Effects and Environmental Fate; Biotechnology; Pesticides*, 29–33.
Patent	Kuhn, K. J., Wehner, W., Zinke, H. (2000) Stabilizer combination for chlorine-containing polymers. US Patent number 6 013 703.
Map	Swansea and The Gower (1974) Ordnance Survey Sheet 159, 1:50 000, First Series. Director General of the Ordnance Survey, Southampton.
Consulting report Include **name of consulting firm**, **contract number** and **for whom** the report was prepared.	Wylie Stream Intake Feasibility Report (2012). James Consultants Ltd., Contract TKA 97/101. Prepared for Middletown Central Electricity Generation.

Undated documents Put (*undated*) where the date is normally placed.	Predicting Traffic Accidents from Roadway Elements on Urban Extensions of State Highways (undated). Bulletin 201, Welsh Highway Research Board.
No author, undated e.g. fact/data sheet, small brochure	Twintex TPP fact sheet (undated). Verdex International S.A.
Video or audio cassette State whether a CD, or video or audio cassette.	Frozen Planet (2011). BBC Natural History Unit Production. Video-cassette.
When none of the above applies	Do what you can to cite enough information to make the source traceable. If there is an author: cite it first. If there is no author, first cite its title (if any), then other relevant details such as the organisation that produced it and any reference number.

Electronic sources

A few notes of caution if you are doing the usual thing of Googling to find sources you can quote in your documents. When you follow a source up, be sure that any source you cite in your report is valid. Academic staff members tend not to regard as credible any citation that does not come from a reputable online source such as a respected institution or organization.

Wikipedia: Don't quote Wikipedia as a source. To quote Wikipedia itself (http://en.wikipedia.org/wiki/Wikipedia:Academic_use):

> *Wikipedia is not considered a credible source. Wikipedia is increasingly used by people in the academic community, from first-year students to professors, as an easily accessible tertiary source for information about anything and everything. However, citation of Wikipedia in research papers may not be considered acceptable, because Wikipedia is not considered a credible source. This is especially true considering anyone can edit the information given at any time.*

However, many Wikipedia articles have references at the end of them. These are worth following up.

Web pages

Conventions for citing web pages are simple. You need to state the following in this sequence:

- **If the site has a stated author:**
  Author (*Family name, followed by initials of given names*). Title of the web page (*in italics*). Retrieved (*date*) from (*URL*).
  Example:
  Siano, D. *The algorithmic beauty of the trebuchet*. Retrieved December 14, 2012 from http://www.algobeautytreb.com/

- **If the site has no stated author:**
  Title of the web page (*in italics*). If possible, the authority under which it appears (this will give credibility to your use of the source). Retrieved (*date*) from (*URL*). The whole URL should be cited, even if it seems very long.
  Example:
  ***An introduction to stand-alone wind energy systems*. Natural Resources Canada. Retrieved August 11, 2012 from http://canmetenergy-canmetenergie.nrcan-rncan.gc.ca/fichier.php/codectec/En/M27-01-1246E/Intro_WindEnergy_ENG.pdf**

If you are using the author/date system, and you need to cite an unauthored source in the text, use the first few words of the title, e.g. See Section 9, sample texts (*The virtual trebuchet*).

Online conference proceedings; an online journal article; abstracts from databases; discussion lists

Use the same system as for a web page.

Emails

Emails are not usually authoritative enough to cite in reports. List them instead as personal communications (see page 175).

Information obtained from an interview

In the text: Treat it as a normal text citation: either [4] or (Atkins, 2012).
In the *List of References*:
Robson, T.G. Robotic Handling Ltd, (2012) In interview with the author.

Example (Both Systems): Text and Corresponding List of References

Author/date system

Note: You don't have to include this column when writing a referenced document; the notes are here just for your information	**Text**
Electronic sources: four with cited authors, one with no cited author.	The recent upsurge of interest in the mechanical efficiency of medieval hurling devices has resulted in their use as student construction projects in engineering (O'Connor, 1994). There is also a wealth of web-based material: for instance, graphics and information (Miners, 2012), desktop models (Toms, 2012), and computer simulations of trebuchets (Siano, 2012; *The virtual trebuchet*, 2012).

Used in ancient times to hurl everything from rocks to plague-ridden carcasses of horses (O'Leary, 1994) and – in a modern four-storey-high reconstruction – dead pigs, Hillman cars and pianos (O'Connor, 1994), the trebuchet relied on the potential energy of a raised weight. Its mechanical efficiency has been compared unfavourably by Gordon (1988) with that of the palintonon, the Greek hurling device, which could hurl 40 kg stone spheres over 400 metres (Hacker, 1968; Marsden, 1969; Soedel and Foley, 1979). This device incorporated huge twisted skeins of tendon, a biomaterial that can be extended reversibly to strains of about 4% (Wainwright et al., 1992). The palintonon used the principle of stored elastic strain energy, the fact that when a material is unloaded after it has been deformed, it returns to its undeformed state due to the release of stored energy (Benham et al., 1996). The motion of the palintonon (Hart, 1982) and that of its Roman equivalent, the onager (Hart and Lewis, 1986), has been analysed by use of the energy principle applied to the finite torsion of elastic cylinders.

Repeat of a previously cited reference

Author mentioned in text

Three references in a series, placed in chronological order, separated by semicolons

An 'et al.' reference – more than two authors

Precise placing of references in the text; one referring to the palintonon, and another to the onager.

Note: Sources are listed in alphabetical order of first author's surname.	***List of References***
Book. Note publisher, place of publication (Harlow) and relevant page(s)	Benham, P.P., Crawford, R.J. and Armstrong, C.G. (1996) *Mechanics of Engineering Materials.* Second edition. Longman, Harlow, page 67.
Book	Gordon, J.E. (1981) *Structures or Why Things Don't Fall Down.* Penguin, Harmondsworth, pages 78–89.
Chapter in book. The book is Volume 9 of a series called Technology and Culture. An 'In:' reference.	Hacker, B.C. (1968) 'Greek catapults and catapult technology: science, technology and war in the ancient world.' In: *Technology and Culture*, **9**, pages 34–50.
Paper in journal	Hart, V.G. (1982) The law of the Greek catapult. *Bull. Inst. Math. Appl.*, **18**, 58–68.
Paper in journal	Hart, V.G. and Lewis, M.J.T. (1986) Mechanics of the onager. *J. Eng. Math.*, **20**, 345–365.
Book	Marsden, E.W. (1969) *Greek and Roman Artillery.* Clarendon Press, Oxford, pages 86–98.
Electronic source with cited author	Miners, R. *The Grey Company Trebuchet Page.* Retrieved May 17, 2012 from http://members.iinet.net.au/~rmine/gctrebs.html

Article in journal, no volume number	O'Connor, L. (1994) Building a better trebuchet. *Mechanical Engineering*, January, 66–69.
Editorial in journal	O'Leary, J. (1994) Reversing the siege mentality. *Mechanical Engineering*, January, 4.
Web page with cited author	Siano, D. *The algorithmic beauty of the trebuchet*. Retrieved August 17, 2012 from http://www.algobeautytreb.com/
Article in magazine	Soedel, W. and Foley, V. (1979) Ancient catapults. *Scientific American*, **240**, 150–160.
Web page, no cited author	*The Virtual Trebuchet*. Retrieved August 18, 2012 from http://heim.ifi.uio.no/~oddharry/blide/vtreb.html
Web page with cited author	Toms, R. *Trebuchet.com*. Retrieved August 21, 2012 from http://www.trebuchet.com/
More than two authors. An 'et al.' reference in the text	Wainwright, S.A., Biggs, W.D., Currey, J.D. and Gosline, J.M. (1992) *Mechanical Design in Organisms*. Second edition. Longman, Harlow. Page 83.

Numbering system

Note: You don't have to include this column when writing a referenced document; the notes are here just for your information.	The recent upsurge of interest in the mechanical efficiency of medieval hurling devices has resulted in their use as subjects for student construction projects in engineering [1]. There is also a wealth of web-based material: for instance, graphics and information [2], applications such as desktop models [3], and computer simulations of a trebuchet [4, 5].
A second reference to Source Number 1. Note: It is not assigned a new number Author mentioned in text Three references in a series, separated by commas Precise placing of references in the text; one referring to the palintonon, and another to the onager.	Used in ancient times to hurl everything from rocks to plague-ridden carcasses of horses [5] and, in a modern four-storey-high reconstruction, dead pigs, Hillman cars and pianos [1], the trebuchet relied on the potential energy of a raised weight. Its mechanical efficiency has been compared unfavourably by Gordon [6] with that of the palintonon, the Greek hurling device, which could hurl 40 kg stone spheres over 400 metres [7, 8, 9]. This device incorporated huge twisted skeins of tendon, a biomaterial that can be extended reversibly to strains of about 4% [10]. The palintonon utilised the principle of stored elastic strain energy – the fact that when a material is unloaded after it has been deformed, it returns to its undeformed state due to the release of stored energy [11]. The motion of the palintonon [12] and that of its Roman equivalent, the onager [13], has been analysed by use of the energy principle applied to the finite torsion of elastic cylinders.

Note: Sources are listed by number in the order in which they appear in the text of the document.	***List of References***
1: Article in journal, no volume number	**1** O'Connor, L. (1994) Building a better trebuchet. *Mechanical Engineering*, January, 66–69.
	2 Miners, R. *The Grey Company Trebuchet Page.* Retrieved May 17, 2012 from http://members.iinet.net.au/~rmine/gctrebs.html
2, 3, 4: electronic sources, each with a cited author	**3** Toms, R. *Trebuchet.com.* Retrieved May 17, 2012 from http://www.trebuchet.com/
	4 Siano, D. *The algorithmic beauty of the trebuchet.* Retrieved May 17, 2012 from http://www.algobeautytreb.com/
5: Electronic source, with no cited author	**5** *The Virtual Trebuchet.* Retrieved February 1, 2012 from http://heim.ifi.uio.no/~oddharry/blide/vtreb.html
6: Editorial in journal	**6** O'Leary, J. (1994) Reversing the siege mentality. *Mechanical Engineering*, January, 4.
7: Book. Note publisher, place of publication and relevant page number(s).	**7** Gordon, J.E. (1981) *Structures or Why Things Don't Fall Down.* Penguin, Harmondsworth, pages 78–89.
8: Article in magazine	**8** Soedel, W. and Foley, V. (1979) Ancient catapults. *Scientific American*, **240**, 150–160.
9: Chapter in book	**9** Hacker, B.C. (1968) 'Greek catapults and catapult technology: science, technology and war in the ancient world.' In: *Technology and Culture*, **9**, pages 34–50.
10: Book	**10** Marsden, E.W. (1969) *Greek and Roman Artillery*. Clarendon Press, Oxford, pages 86–98.
11: Book with four authors	**11** Wainwright, S.A., Biggs, W.D., Currey, J.D. and Gosline, J.M. (1992) *Mechanical Design in Organisms*. Second edition. Longman, Harlow. Page 83.
12: Book with three authors	**12** Benham, P.P., Crawford, R.J. and Armstrong, C.G. (1996) *Mechanics of Engineering Materials.* Second edition. Longman, Harlow, page 67.
13: Paper in journal	**13** Hart, V.G. (1982) The law of the Greek catapult. *Bull. Inst. Math. Appl.*, **18**, 58–68.
14: Paper in journal	**14** Hart, V.G. and Lewis, M.J.T. (1986) Mechanics of the onager. *J. Eng. Math.*, **20**, 345–365.

Using direct quotations

You may occasionally need to quote word for word from another source. This may be particularly so in essays where you are writing about contentious issues and feel that the exact words are relevant to your discussion. These quotations need to be enclosed in quotation marks.

Do not make the naive mistake of thinking you can avoid plagiarism by including from another source large amounts of word-for-word text contained between quotation marks. This convention applies only to direct quotations that are necessary to your argument.

Conventions for direct quotations

A direct quotation Enclose it in quotation marks ("…").	According to Huxley, "science is nothing but trained and organised common sense".
Where very slight changes are needed to a quotation so that it fits into your prose:	Pratchett (1993) has noted that '[she] lived in the kind of poverty that was only available to the very rich, a poverty approached from the other side'. (The original quote: 'Sybil Ramkin lived in…')
For example, a capital letter may need to be changed to lowercase, or a noun substituted for a pronoun, or a noun or phrase inserted, so that it makes more sense. These changes are indicated by square brackets [ ].	Gould (1985) has stated that "[t]he history [of human races] is largely a tale of division – an account of barriers and ranks erected to maintain the power and hegemony of those on top". (*The original quote: "The history is largely…"*.)
Where part of a quote needs to be omitted because it is irrelevant to your document:	Watson (1968) has stated "… I had chosen the wrong tautomeric forms of guanine and thymine".
Use three dots to show the omission. It is important that the sense of a quotation is not altered by the omission.	

Compiling a bibliography

The conventions used for compiling a *Bibliography* are given here:

- Use the same method of writing out each item as you would for a *List of References* section (see pages 177–182).
- The items are listed in alphabetical order according to the first author's surname, or the title of the reference if the author is unknown.
- The list is not numbered.
- It is common practice to indent each line of a reference after the first. Use the hanging indent function on a word processor.

Example:

Bibliography

Benham, P.P., Crawford, R.J. and Armstrong, C.G. (1996) *Mechanics of Engineering Materials*. Second edition. Longman, Harlow, page 67.

Siano, D. *The algorithmic beauty of the trebuchet*. Retrieved August 17, 2012 from http://www.algobeautytreb.com/

The Virtual Trebuchet. Retrieved August 18, 2012 from http://heim.ifi.uio.no/~oddharry/blide/vtreb.html

Toms, R. *Trebuchet.com*. Retrieved August 21, 2012 from http://www.trebuchet.com/

Wainwright, S.A., Biggs, W.D., Currey, J.D. and Gosline, J.M. (1992) *Mechanical Design in Organisms*. Second edition. Longman, Harlow.

Plagiarism and how to avoid it

Plagiarism is literary theft. It occurs when you do the following:

- Copy material from books, the Web and other sources.
- Quote any piece of information that is not common knowledge.
- Use another person's theory or opinion without crediting that person.
- Lightly paraphrase (slightly reword) another person's written or spoken words.

Examples

1. When you make a statement that your reader needs to know is valid, you must cite the source:
   Wind energy is the world's fastest growing energy source.
   Who says so? Can your reader trust this information that you've given? Answer: The reader can only trust this information when you state from which source you got your material. You must therefore cite the source in your document and list it in your *References* section.
2. When you block-copy material from the Web or books without citing the source. Many students have block-copied web material for assignments, believing that their assessor will not notice it, or in ignorance of the fact that it shouldn't be done. By doing this, you are implying that the material is your own. You need to (1) rewrite it (that is, paraphrase it) and (2) cite the source. Be careful: You should *not* just alter a few words here and there. To avoid plagiarism, you need to substantially rewrite it in your own words.

The following passage has been block-copied from the web page of the US Department of Energy's Wind Energy Program web page (http://www.eren.doe.gov/wind/web.html).

> *What causes the wind to blow? Wind is a form of solar energy. Winds are caused by the uneven heating of the atmosphere by the sun, the irregularities of the earth's surface, and rotation of the earth. Wind flow patterns are modified by the earth's terrain, bodies of water, and vegetative cover. This wind flow, or motion energy, when "harvested" by modern wind turbines can be used to generate electricity.*

Here's an unacceptable paraphrase that is **plagiarism:**

Wind flow, or motion energy, when "harvested" by modern wind turbines can be used to generate electricity. What makes the wind blow? Wind, a form of solar energy, is caused by the uneven heating of the earth's atmosphere by the sun, the earth's irregular surface, and its rotation. Wind flow patterns are modified by the earth's terrain, seas or large lakes, and vegetative cover.

Why is this passage plagiarism?

It is considered to be plagiarism for two reasons:

1. The writer has changed around only a few words and phrases, or changed the order of the original paragraph's sentences.
2. The writer has failed to cite a source for any of the facts.

If you do either or both of these things, you are plagiarising.

Here is an acceptable paraphrase:

Wind has three causes: uneven heating of the earth's atmosphere by the sun, the topography of the earth's surface, and the earth's rotation. Wind can be used to generate electricity; however, because flow patterns are modified by topography, oceans and the amount of plant cover, wind flow can be very variable [12].

Why is this passage acceptable?

This is acceptable paraphrasing because the writer does the following:

- Accurately transmits the original information in a substantially rewritten form.
- Cites the source of the information.

3. **Downloading diagrams from the Web.** Make sure that each of your diagrams in your report has its own figure number, a title of your own making (even if the diagram comes off the Web with its own title), and that the source is cited at the end of the title. For example:

   Figure 4 Cost of wind-generated electricity, 1980–2005 [15]

   For the citation conventions when copying, adapting or redrawing diagrams, see page 175, in this chapter.

4. **If you need to quote something word for word**, use the conventions for quotation marks (see page 186, in this chapter), and cite the source.

How to avoid plagiarism

- Don't copy word-for-word material out of books, off the Web or from other sources.
- Paraphrase the material (substantially rewrite it and express it differently), and cite the source.
- Diagrams: devise your own title for the diagram. Cite the source at the end of the title.

Common mistakes

1. Citing a reference in the text and leaving it out of the *List of References.*
2. Citing a reference in the *List of References* and making no mention of it in the text.
3. The date of the text citation does not correspond with that of the listing in the *List of References.*

These three mistakes above tend to be regarded as unforgivable by most university staff.

In the *List of References* section:

1. Using non-standard abbreviations for a journal.
2. Giving insufficient details in the *References* section; in particular, omitting the publisher and place of publication of a book, omitting the date.
3. Using inconsistent formatting.
4. Using incorrect volume and page numbers.
5. Giving unobtainable references.

Checklist: Referencing

- ☐ References are needed in the following situations:
  - ☐ You cite factual material from the literature.
  - ☐ You quote directly from another work.
- ☐ Decide whether you need a *References* section or a *Bibliography*. Most institutions need a *References* section.
- ☐ ***References* section:** the two main systems – there are minor variations – of citing references are listed here:
  1. Author/date system
  2. Numbering system

Use the one recommended by the institution or journal for which you are writing the document

Use one system or the other consistently. Never use a mixture of the two.

- ☐ Whichever of these systems you use, you must have a section called ***List of References*** at the end of your document:
- ☐ In the **author-date system**, the sources are listed alphabetically by the surname of the first author.
- ☐ In the **numbering system**, they are listed sequentially according to the number given them in the text.
- ☐ A reference should appear only once in the *References* section.
- ☐ Make sure each reference is formatted consistently and accurately.
- ☐ **In the text**, cite each reference according to the conventions of the system you are using.
- ☐ For a **Bibliography**, use the same conventions for writing out each of the full references. Then list them alphabetically.

Then check the following:

- ☐ For each one of your text citations, is there a corresponding reference in the *References* section? And vice versa?
- ☐ Does the date of the text citation match the date in the full reference in the *References* section?
- ☐ Are all of the references in the *References* section formatted consistently?
- ☐ Are all of the necessary details there?

Avoiding Plagiarism

- ☐ Have you avoided copying word-for-word material out of books, off the Web or from other sources?
- ☐ Have you substantially rewritten the material and cited the source?
- ☐ Does each figure have its own figure number and its own title? Is the source cited at the end of the title?

16 Conventions Used in Scientific and Technical Writing

This chapter covers conventions for the following:

- Where to place the titles of figures and tables
- Using numbers
  - Numbering of illustrations, sections, pages, appendices, equations
  - Writing numbers in the text
- Referring in the text to figures, tables, chapters, table rows or columns, pages
- Equations: formatting in the text
- Written style for headings
- SI units
- Genus and species names
- Checklists

Obeying the conventions may sound tedious. However, there are many standard conventions of scientific writing and to maximise your marks for a report, you'll need to use them. Staff assessors will expect these conventions to be used and are likely to mark you down if you don't. It will be particularly important for a major report in the later years of your degree.

This chapter is a collection of the main conventions used. Some of them are also briefly referred to at the appropriate part of the various chapters.

Where to Place the Titles of Tables and Figures

The table number and title is placed *above* a table.
The figure number and title is placed *below* a figure.

This is just one of those strange conventions. However, some graphing programs don't conform with it.

Common mistake

Table headings placed under the table.

Writing for Science and Engineering.
DOI: http://dx.doi.org/10.1016/B978-0-08-098285-4.00016-9

Using Numbers

Numbering of Illustrations, sections, pages, appendices and equations

Numbering of illustrations

- **Every illustration (figure or table) in a document *must* have the following:**
  - **A number**
  - **A title**
  - **And be referred to at an appropriate place in the text.** See **Illustrations** Chapter 2, page 44, *The Core Chapter*, for other information about requirements.

- **There should be two numbering series for illustrations:**
  1. **One for all the figures**, i.e. everything that isn't a table – graphs, maps, line drawings, schematics, etc.
  2. **Another series for the tables.**

  This means that there will be Figure 1, Figure 2... etc., **and** Table 1, Table 2... etc.

Conventions for figure numbering

All your figures – this includes graphs, line drawings, maps, photographs and other types of illustrations – should be labelled as one series. Each one of the series is then labelled Figure 1, Figure 2, etc.

Example: If the first of your illustrations is a map, the second a graph and the third a line drawing, call the map Figure 1, the graph Figure 2 and the line drawing Figure 3.

Common mistake

To distinguish between the different types of illustrations and to call your graphs Graph 1, Graph 2..., your maps Map 1, Map 2... and your line drawings Figure 1, Figure 2....

Variation on This: Numbering According to the Section Number

When your report has numbered sections (see *Numbering of Sections*, page 193), your illustrations can be numbered according to the section numbers:

First illustration	A map in Section 2	Figure 2.1
Second illustration	A graph in Section 2	Figure 2.2
Third illustration	A line drawing in Section 4	Figure 4.1
Fourth illustration	A graph in Section 5	Figure 5.1

Do not try to number your illustrations according to the sub-section number. You will end up with figure numbers such as Figure 3.1.2.1 (the first figure in

Section 3.1.2), which is clumsy and confusing. Number them in accordance with the main section only.

Table numbers

The numbering series for your tables is completely independent of the series for your figures.

Number each table Table 1, Table 2… in sequence.

Variation on this

In the same way as for figures, tables can be numbered according to the section of your report in which they occur.

First table	The first one in Section 2	Table 2.1
Second table	The second one in Section 2	Table 2.2
Third table	The first one in Section 5	Table 5.1

Numbering and captions of tables and figures in the *appendices*

Tables and figures in *Appendices* do not belong to the two series in the main body of the document. They are labelled as two separate series in their own right, according to the numbering of the Appendix.

Figure 1-1 (Figure 1 in Appendix 1), Figure 1-2 (Figure 2 in Appendix 1) etc.
or
Figure A-1 (Figure 1 in Appendix A), Figure A-2 (Figure 2 in Appendix A) etc.
or
Figure 2, Appendix 1, etc.

Numbering of sections of a report

This section describes the conventions for the decimal point numbering system for numbering sections of a document, and their associated sub-headings and sub-sub-headings.

The main sections are given Arabic numerals. The sub-sections are shown by putting a decimal point after the section number and another Arabic numeral:

1.0 Title of first main section
    1.1 First sub-heading
    1.2 Second sub-heading
2.0 Title of second main section
    2.1 First sub-heading
    2.2 Second sub-heading
        2.2.1 First division in the second sub-heading
        2.2.2 Second division in the second sub-heading
        2.2.3 Third division in the second sub-heading
    2.3 Third sub-heading
3.0 Title of third main section

For the formatting conventions of a *Contents Page*, see *Table of Contents*, Chapter 2, *The Core Chapter*, page 23.

Numbering of pages

The conventions associated with page numbering in a document are the following:

- All of the preliminary pages, i.e. those before the *Introduction* (*Title page*, *Abstract*, *Acknowledgements*, *Table of Contents*, *List of Illustrations*, *Glossary of Terms and Abbreviations*, etc.) are assigned lowercase Roman numerals (i, ii, iii, iv, v, etc.).
- The first page that is counted is the Title Page, but it is not labelled as such; it is left blank.
- Each of the other preliminary pages (starting at page ii) is labelled with its number.
- Page 1 is the first page of the *Introduction*.
- After page 1, the page numbering is continuous.

Any large scientific document such as a thesis or a major report will be expected to conform to this numbering system.

> **Common mistake**
>
> To number the *Abstract* page as page 1 and continue from there.

Numbering of appendices

Appendices can be named either:

Appendix 1, Appendix 2, Appendix 3, etc.
or
Appendix A, Appendix B, Appendix C, etc.

The page numbers of the *Appendices* are usually separate from those of the main body of the document and are related to the numbering of the *Appendix*. For example:

Page 1-1, 1-2. 1-3 etc. *or* page A-1, A-2. A-3 etc.

However, word processors can be uncooperative in producing this sort of numbering, so many assessors might accept a serial numbering for all of the appendix pages, or even a continuation of the page numbering of the main body of the report. Check with the assessor.

Numbering of equations

Equations should be numbered consecutively throughout the report, either as a continuous series (1, 2, 3, 4, etc.) or be related to the section number (e.g. Equation 3.2 would be Equation 2 in Section 3 of the report, etc.).

Writing numbers in the text

These are the conventions for dealing with numbers in scientific and technical writing:

Summary Table

Numbers	**Rule/Convention**	**Example**
Measured quantities	Figures	6 tonnes, 3 amps
Counted numbers		
One to ten	Words	… in five areas
More than ten	Figures	… in 11 areas
Number at the beginning of a sentence	Words	Eleven samples were taken.
Ordinal numbers	Same as for counted numbers (above)	(first, second, third… 11th…)
A series of numbers above and below 10	Figures	… over periods of 3, 6 and 12 hours
Percentages	Figures	… that 8% of the samples…
Fractions	Words	… one-fifth of the bait was taken *(but better expressed as a percentage)*
Dates and times	Figures	… on 8 October … at 8.30 am (or 08:30)
Reference in the text to figures and tables	Figures	Figure 3 shows that… … (Table 2)

Referring in the Text to Figures, Tables, Chapters, Rows or Columns of Tables, Pages

		Examples
Figures, tables, chapters, sections	Use initial capitals when referring to specific figures, tables, chapters or sections.	… is shown in Figure 3. … as given in Table 2. … is described in Chapter 6. … is analysed in Section 5.
Rows or columns of tables, pages	Do not use initial capitals when referring to rows or columns of tables, or to pages.	… as given in row 2 of Table 12. … as given in column 3 of Table 12. … is given in Section 4, pages 38–41.

Equations: Formatting in the Text

Equations should be centred with the equation numbers in round brackets and right justified (see above for equation numbering). Leave about one line of space both above and below the equation. For instance:

$$y = ax + \cos x + \beta \tag{1}$$

There are minor variations in styles of formatting equations. The following shows a good general style:

Notes	Text of the document
Equation is centred	The value of the shear stress at a distance *r* from the axis is given by
Equation number in brackets is tabbed to the right margin. This is equation number 5 in Section 3 of the report.	$$\tau = Gr\frac{\mathrm{d}\phi}{\mathrm{d}x} \qquad (3.5)$$
In the text refer to the equation as either 'Eq. (*equation number*)' *or* 'equation (*equation number*)'. Be consistent in your use of one or the other throughout your text.	Eq. (3.5) shows that the shear stress acting on the circular cross-section is linear in the radius *r*.

For a sequence of equations in which the left-hand side is unchanged. Align the = symbol in each line.

$$\begin{aligned} u(x) &= -\frac{q_0}{AE}\int_0^x (x-\xi)\mathrm{d}\xi + \frac{C_1 x}{AE} \\ &= -\frac{q_0 x^2}{2AE} + \frac{C_1 x}{AE} \end{aligned}$$

For continued expressions in which the left side is long. Align the = symbol with the first operator in the first line.

$$\begin{aligned} &[(a_1 + ia_2) + (a_{11}s_1 + a_{21}s_2)]/[(b_1 + ib_2) + (b_{11}s_1 + b_{21}s_2)] \\ &\quad = f(x)g(y) + \ldots \end{aligned}$$

For expressions in which the right-hand side is long: Align the continuing operator with the first term to the right of the = symbol.

$$\begin{aligned} V(x) = &- P\langle x\rangle^0 + P(x-a)^0 \\ &+ P\langle x-(L-a)\rangle^0 - P\langle x-L\rangle^0 + C_1 \end{aligned}$$

Built-up fractions should be avoided in text. Instead, use solidus fractions $1/(x+y)$.

Written Style for Headings

Headings and Sub-Headings of Sections in the Text and the Table of Contents

Initial capitals can be used to distinguish between the different levels of headings. For example:

Top level headings use the format known as Title Case (main words are capitalised):

Example: **Results and Discussion**

Next heading level down: sentence case (first word is capitalised):

Example: **Implementation of scheduling**

***General convention*: Capitalise the initial letters of the 'main' words in the titles, i.e. the words other than small words such as articles, prepositions and conjunctions (Table 16.1).**

Table 16.1 Common Words That Do Not Have Initial Capitals, Unless They are the First Word in the Title

Articles	the a
Prepositions	across by for to up down of
Conjunctions	and but so since because for although
Coordinators	if…then both…and either…or neither…nor whether…or

Use of SI Units

The International System of Units (SI) should always be used throughout scientific and technical documents. If you don't do this, or if you use a mixture of units, your reports are likely to be severely penalised.

The SI units most commonly used in student scientific and technical reports are given in Appendix 1, *SI Units and Their Abbreviations*, page 257.

For greater detail, use the authoritative online source:

International System of Units (SI): http://physics.nist.gov/cuu/Units/units.html

Genus and Species Names

The conventions for the more simple aspects are given here. For further detail, use a style manual such as *Scientific Style and Format: The CSE Manual for Authors, Editors and Publishers.* 7th edition. Council of Science Editors, 2006.

The name of a species is in two parts, consisting of two Latin names: a genus name and a species epithet, e.g. *Panthera tigris*; *Rosa acicularis*.

Be aware that the conventions governing sub-species differ between zoology and botany.

- A zoological sub-species is in three parts: *Panthera tigris sumatrae*.
- A botanical sub-species requires 'ssp.' or 'subsp.' before the sub-specific name: *Rosa carolina* subsp. *subserrulata*.
- The initial letter of the genus name is capitalised.
- The initial letter of the species epithet is always written in lowercase.

Common mistake

To capitalise the species epithet, e.g. *Panthera Tigris*.

- The whole name is italicised (or underlined if you are writing by hand).
- A genus name should always be followed by a species epithet or, if the species is unknown, by 'species', 'sp' (singular) or 'spp' (plural), none of which are italicised, e.g. *Rosa* sp.
- A genus name should be spelled out on first mention in the text. Thereafter, it can be abbreviated to the initial letter followed by a full-stop and the species epithet, e.g. *P. tigris*; *R. carolina*.
- A variety is written as follows: *Rosa acicularis* var. Alba, or *R. acicularis* var. Alba.

Checklist

Make sure you use all of the relevant conventions consistently and correctly.

17 Revising

This chapter covers the efficient multi-pass process of revising a document, in which different aspects are revised at each pass.

Before You Revise

Allow Plenty of Time for Rewriting

The final stages of the writing process – revising, retyping or rewriting, inserting illustrations etc. – always take much, *much* longer than expected. Allow ample time for them in your planning schedule; otherwise, you'll find yourself having to submit an early draft.

Avoid 'Memory Reading'

After writing something, most people become blind to what they have written. When you have read the document a number of times, you become too familiar with the text. You will tend to read from memory and miss some of the mistakes. There are two ways of overcoming this:

- **Ask someone else to read it.**
- **Stand back from it for as long as possible**. Most undergraduates don't have the time; however, try to put it aside for as long as you can. Even half an hour helps; a day or more is much better. Mistakes of organisation and style then leap out at you.

Multi-Pass Revising

Advice on how to edit efficiently says that you will need to edit at different levels by making several passes through the document, looking at only certain aspects in each pass. This sounds very time-consuming, but it's been shown to take less time – and to be more effective – than trying to do it all in one pass.

Writing for Science and Engineering.
DOI: http://dx.doi.org/10.1016/B978-0-08-098285-4.00017-0

However, books suggest two different schemes: one scheme goes from overall organisation to the minute details; the other scheme recommends the reverse. Choose the one you prefer!

Level	*Scheme 1*	*Scheme 2*
	Going from details to overall organisation	*Going from overall organisation to details*
Level 1	• Style • Spelling, typos • Grammar • Punctuation	• Overall organisation
Level 2	• Paragraph and sentence length and structure • Verbiage (too many words) • Precise word choice	• Style • Spelling, typos • Grammar • Punctuation
Level 3	• Overall organisation • Format • Appearance	• Formatting
Level 4	Document integrity; matching of the following: • Section numbers to Contents Page • Page numbers in text and Contents Page • Text references and figure numbers	

Note: By using word-processing software in Outline mode and the cross-referencing functions, the stages involving overall organisation and document integrity are made very much easier (see page 11 for brief information on Outline mode). These functions will produce an automatic *Table of Contents* and correlate figure numbers with text references.

Aspects to Check: Organisation, Style, Formatting, Document Integrity

Organisation

Examine your assignment critically in an *overall* way. Look only for errors of organisation.

a. Have you followed your plan?
b. Is the structure logical?
c. Should this section come before that one, that paragraph before this one?
d. Is the information coherent?

Style, Grammar, Sentence and Paragraph Length, Punctuation

At this stage, polish up details of style. For help, see Chapter 18: *Problems of Style*, page 207.

Paragraphs

- Are the breaks between the paragraphs logical?

Sentences

- Are you writing in 'real' sentences?
- Are your sentences too long?
- Are they bitty? Do they need linking?

Verbs

- Are you using the distorted passive or other lifeless verbs?
- Are you using the right tense?
- Is there subject–verb agreement?

Words

- Have you spellchecked?
- Are you using pompous words where short ones would be better?
- Are you using jargon or clichés?
- Are you using *I* or *We* too often?
- Is your language gender-neutral?
- Are there colloquialisms or contractions?
- Are you using the right word? (*affect/effect, led/lead, lose/loose, etc.*)
- Are you writing numbers correctly? (*Ten or 10?*)

Punctuation

- Are your commas and full stops effective?
- Are the apostrophes used correctly?

Formatting

If you are writing a journal paper, make sure you follow the journal's *Instructions to Authors*, which will give full formatting instructions.

If you are writing any type of document other than a journal paper: Here are guidelines for fonts, paragraph formatting, justification, packing density of the text and page breaks.

An assignment should be visually strong, not dauntingly dull looking. But make sure that each formatting decision you make is for a good reason. If you just play with it, you'll end up with a document that looks disordered.

If you are word processing your document, you have many options:

a. Fonts. Use only the plain fonts; your reader will strongly dislike the more elaborate fonts. One font per document looks professional: either a plain serif or a plain sans serif. Use a maximum of two in one document; a plain serif font for the text and a plain sans serif for headings can look good. Too many different fonts look messy.

b. Paragraphs. There are two methods for formatting the first line of a paragraph:

1. To leave a blank space between each paragraph and the next (press *Enter* once)

2. To indent the first line of each paragraph by no less than five characters

Common mistake

No space, no indentation (or a slight indentation of only one or two characters). This may make it very difficult to distinguish one paragraph from the next.

c. **Justification of the text.** Justified right-hand margins can make a document look professionally tidy, but there may be problems with the appearance of the spacing between words.

d. **Packing density of the text.** Information is more readily absorbed if it's not too dense on the page:
   - **Wide margins.**
   - **Bullet points in the text of reports.** This is a useful method for listing within the text. However, too much use of bullet-pointing can reduce a document to chaos.
   - **Indented left-hand margins**, for instance when using bullet points or quotations. However, avoid indenting so many times that the text is squeezed into the right-hand side of the page.

e. **The amount of space between lines.** Be careful: Your assessor will think you are trying to hide lack of content if the line spacing is too wide and/or the font is too large (larger than 12 point serif).

f. **Effective page breaks.** Avoid the following bad breaks:

   A heading at the bottom of the page (there should be at least two lines of text following a heading)
   A short line (a widow) at the top of the page
   A table that is cut in two by a page break
   A page that ends with a hyphenated word

Document Integrity

If possible, use the word-processing functions that will enable the following automatic functions:

- Numbering of section headings
- Creating a *Table of Contents*
- Cross-referencing of figure and table numbers and their text references
- Cross-referencing your text citations with the sources in your *List of References.*

If you cannot do this, then these common problems may occur, and you will need to manually check whether there are discrepancies in the following:

- The numbering of section headings
- Referring to figures in the text
- Referencing
- The *Table of Contents* and the corresponding page numbers
- The illustrations

***Section headings:** Is the numbering of each heading and its sub-headings consistent?*

***Illustrations:** Is each illustration:*

- Numbered?
- Titled?
- Adequately labelled?
- Correctly referred to in the text?

References: *For each reference:*

- For each text citation, is there the corresponding reference in the *List of References* section? And *vice versa*?
- Does the date of the text citation match the date in the full reference in the *List of References* section?
- Are all the references in the *List of References* section formatted consistently?
- Are all the necessary details there?

Table of Contents (see Chapter 2, *The Core Chapter*, page 23)

- Does the wording of headings match up with the headings in the text?
- Is the numbering of each heading and its sub-headings consistent?
- Is the formatting (indenting) of the Contents page consistent?
- Are the page numbers correct?

Section 4

Writing Style

18 Problems of Style: Recognising and Correcting Them

Scientific and technical writing is about clarity and conciseness. This chapter doesn't try to give comprehensive guidelines on stylistic elegance. Instead, using simple terms and avoiding classical grammar as far as possible, it deals with some of the straightforward, frequently asked questions about style. It covers the following:

- **Punctuation**
  - The apostrophe
  - Commas, semicolons and colons
  - Exclamation marks: don't use them
  - Rhetorical questions: don't use them
- **Words**
  - Spelling: check it
  - Making plurals/irregular plurals
  - Noun trains
  - Pairs of words that are often mixed up
  - Jargon phrases to avoid
  - Use small words, not pompous ones
  - The split infinitive
  - Sentence length
  - Paragraph length
- **Verbs and vivid language**
  - The voice: active/passive/distorted passive
  - Lifeless verbs
  - Excessive use of nouns instead of verbs
  - Subject/verb agreement
  - Using the correct tense/form of the verb
- **Incomplete sentences: recognising and correcting them**
- **When English is a foreign language**

Writing for Science and Engineering.
DOI: http://dx.doi.org/10.1016/B978-0-08-098285-4.00018-2

Punctuation

The apostrophe

There are two areas where an apostrophe is used:

1. **The possessive**: showing who or what something belongs to.
2. **Contractions**: where two words have been informally squashed into one.

The possessive

The apostrophe shows who/what owns something.
Use **'s** if there is only one:
the cell's chromosomes (the chromosomes in one cell)
Use **s'** if there are more than one:
the cells' chromosomes (the chromosomes in a number of cells)
BUT, there is no apostrophe in **yours** (e.g. the book is yours), **hers, ours, theirs** or in **its** (e.g. the cell and its chromosomes, *see also next section*).

Contractions: Don't use them

> *Note: This book uses contractions throughout, but it's not an example of formal writing.*

The apostrophe is used in a contraction to show that two words have been informally pushed together. Because contractions are informal, they shouldn't be used in the types of writing covered in this book.

The main contractions that cause confusion are the following:

1. **it's**
2. **who's**
3. everything ending in … **n't** (wouldn't, hadn't, etc.)

It's/Its

The mixing up of these two happens all the time, yet it's very easy to understand the difference.

It's is the contracted, colloquial way to write **it is** or, less often, **it has**. Therefore: **Never use *it's* in any formal writing. It's colloquial. It can *only* mean it is.**

Wrong	*Right*
Because of overuse, the land has lost it's nutrients.	**Because of overuse, the land has lost its nutrients**

Putting it Right: it's/its

- **Never** write **it's** in formal writing. **It's = it is.**

For anything other than informal letters or notes, you will need the **its** form – with no exceptions.

A good way to check:
Read it aloud to yourself, saying every **it's** as **it is**. Does **it is** make sense? If so, write it. If not, write **its**.

Whose/Who's

This case is very like *it's/its*. **Who's** is a colloquial form of **who is** or **who has**.

Wrong	*Right*
Mr. Smith, who's responsibility is the monitoring of the outfall, says that …	**Mr. Smith, whose responsibility is the monitoring of the outfall, says that … (Does it mean *who is*? No: so write *whose*.)**
Mr. Smith, who's responsible for monitoring the outfall, says that …	**Mr. Smith, who is responsible for monitoring the outfall, says that … (Does it mean *who is*? Yes: so write *who is*.)**

Putting it Right

- If you mean **who is** or **who has**, write it.
- All the other times, you'll need **whose**.

Everything ending in …n't

Examples:

shouldn't, mustn't, wouldn't, didn't, can't, hadn't, etc.

In any formal writing, the words should be written out in full:

should not, must not, would not, did not, cannot, had not, etc.

Examples of contractions. ***Incorrect in formal writing***	*Wrong*	*Right*
couldn't	**The valve couldn't be opened.**	could not
wouldn't	**The young birds wouldn't feed.**	would not
wasn't	**The stream wasn't polluted.**	was not
weren't	**The older birds weren't present.**	were not
didn't	**The water didn't contain PCBs.**	did not
shouldn't	**This procedure shouldn't have been followed.**	should not
hadn't	**The company said it hadn't been informed.**	had not

Putting It Right: Summary

Never use contractions in formal writing. Write them out in full. The following are particularly common:

- **it's** = it is
- **who's** = who is
- **words ending in** ...***n't*** **(don't, won't, can't, couldn't, shouldn't, etc.** = do not, will not, cannot, could not, should not, etc.)

ALSO: plurals are not made by adding 's. See *Plurals*, page 213.

Commas, semicolons and colons

(For the use of **quotation marks** in referencing, see Chapter 15: *Referencing*, page 186.)

The following provides very brief guidelines to the main ways in which commas, semicolons and colons are used.

Using a comma

A comma indicates a pause. You can often tell where a comma should be by saying the words to yourself. Commas are generally used in the following places:

***After each item in a series but generally not before the final and*:**

(*Adjectives*) **The river is wide, turbulent and muddy.**

(*Nouns*) **The most common birds on the island are sparrows, chaffinches, thrushes and blackbirds.**

(*Phrases*) **The river mouth is wide, with large shingle banks, extensive sand dunes and a small island.**

To delimit a sub-clause from the main clause in a sentence:

When the engine was run on petrol, the carbon dioxide emissions were higher, which was an indication of improved mixing.

Increasing agriculture will cause an increase in global warming, the reason being that ruminants and paddy fields produce methane.

After an introductory phrase or sub-clause:

Although farmers in this area have reduced their use of pesticides in recent years, there is still local concern about the issue.

By using better management practices, farmers have been able to reduce their use of pesticides.

To delimit material that is not essential to the meaning of the sentence:

The island, although windswept, has a large number of different bird species.

Using a semicolon

- ***Between two closely related independent clauses:***

  (The statement on each side of a semicolon should be able to stand alone as a sentence.)

  The spill caused the level of toxins in the river to rise; as a result, the entire fish population died.

- ***Between items in a list when the items are punctuated by commas:***
  Yesterday I ate muesli, bacon and eggs for breakfast; bread, cheese and pickles for lunch; and fish and chips for dinner.

Using a colon

To introduce a list or series:

1. ***Before bullet points:***
   The following topics will be discussed:
   - **climate change**
   - **volcanic hazards**
   - **violence against women**
   - **poverty**
2. ***Where a list is all strung together, a colon precedes the listed points, and semicolons separate them:***
   The following topics will be discussed: climate change; volcanic hazards; violence against women; and poverty.
   Several features changed significantly during the sampling period: water temperatures decreased; ammonium levels increased to more than 150 ppm; dissolved oxygen fluctuated; and the *p*H rose at one stage to 8.3.

Exclamation marks: Don't use them

Breathless writing and exclamation marks are not appropriate in formal writing. Avoid phrasing such as the following:

The world has a problem with carbon dioxide, of that we can be sure!
There is now the possibility of restoring these sites back to their original (hopefully!) condition.

Rhetorical questions: Don't use them

These are questions asked to produce an effect rather than to gain information. They are usually not appropriate in a technical document. But they are frequently used by inexperienced writers and sound very clumsy.

Wrong: **This study has shown that the nutrient level is low. What can be done about it? It can be remedied by …**

Corrected: **This study has shown that the nutrient level is low. It can be remedied by …**

Words

Spelling: Check it

Never underestimate the effect that bad spelling has on the quality of a piece of writing. Some people don't notice when words are spelled incorrectly; others are irritated because it interrupts the flow of the text. Many assessors fall into the second category.

Use the spellchecker on the word processor, and then proofread it

- **Run everything through a spellchecker as the very final stage before submission.** The critical word here is *final*. Many mistakes creep in when last-minute amendments are done and spellchecking is then omitted in the general haste.
- **Even after spellchecking, never assume that a spellchecked assignment is error-free. Proofread it yet again.** A spellchecker will pass words that you may not have intended – *it* instead of *is*, *an* instead of *on*, etc.

Example of spellchecked nonsense

Here is a constructed example of something that would be passed by a spellchecker but is nonsense. Each word has a maximum of only one mistyped letter.

His technique cam also by applies to the analyses or gold bills. He surface oh a gulf bell hat dimpled an is, ant whet is travels thorough aid, the flop around the bell it smother.

The intended version:

This technique can also be applied to the analysis of golf balls. The surface of a golf ball has dimples on it, and when it travels through air, the flow around the ball is smoother.

Commonly misspelt words in technical writing

Wrong	*Right*
accomodation	accommodation
callibrated	calibrated
comparitive	comparative
consistant	consistent
equillibrium	equilibrium
guage	gauge
heirarchy	hierarchy
intergrate	integrate
proceedure	procedure
recomend/reccomend	recommend
rythm	rhythm
seperate	separate
speciman	specimen
theoritical	theoretical
verses	versus (as in describing a graph)
vise versa (and variations)	vice versa
yeild	yield

Plurals

- Never make a plural – more than one of something – by adding *'s*. Just add *s*. There are only two types of exceptions to this:
  - add -es if the word ends in *-o* or ***-sh*** or *-ss* (*potatoes* or *fishes* or *classes*)
  - add -ies if the word ends in -**y** (*ferries*).
- The plurals of abbreviations or dates don't have apostrophes.

Wrong	*Right*
Many river's	Many rivers
Many valve's	Many valves
Many potato's	Many potatoes
Many fish's	Many fishes
Many pizza's	Many pizzas
Many ferry's	Many ferries
Many class's	Many classes
PCB's	PCBs
The 1990s	The 1990s

Irregular plurals

The following words are commonly used in science and technological writing and have irregular plural forms:

Singular	**Plural**
alga	algae
analysis	analyses
antenna	antennae (biology) antennas (communications engineering)
appendix	appendices
axis	axes
bacterium	bacteria
criterion	criteria
genus	genera
hypothesis	hypotheses
larva	larvae
locus	loci
matrix	matrices
medium	media
nucleus	nuclei
ovum	ova
phenomenon	phenomena
quantum	quanta
radius	radii
species	species
stimulus	stimuli
stratum	strata
symposium	symposia
vertebra	vertebrae
vortex	vortices

Noun trains

These are strings of nouns that are each piled on top of another. They can be found in titles, as a result of an attempt to compress as much information into as few words as possible. The significant information is the part at the end. For example:

The Middleborough Point power station fan floor concrete slab

To rewrite, take the final part, bring it to the beginning, and it will readily sort itself out.

The concrete slab forming the fan floor of the Middleborough Point power station

The rewritten form will inevitably be longer, but it will be easier to read.

This example is an eight-part noun train. It's usually accepted that a three-part noun train is understandable; anything above three parts should be rewritten.

Pairs of words often mixed up

There are pairs of words or expressions that are often muddled. Some of the most common pairs are:

Absorb/adsorb
Affect/effect
Complement/compliment
Imminent/eminent
It is composed of/it comprises
Its/it's
Lead/led
Loose/lose
Passed/past
Principal/principle
Their/there
Whose/who's

Absorb/Adsorb

Absorb

- To take up by chemical or physical action
- The swallowing up or engulfing of something

Adsorb

The process of the adhering of atoms or molecules to exposed surfaces, usually of a solid. **It should be used only when you need this precise meaning.**

Affect/Effect

This is easy when you know how. Focusing on the commonest uses of the two words in most science and technological writing:

affect is a verb, **effect** is a noun (see Appendix 2, *The Parts of Speech and Verb Forms*, for guidelines on these terms).

To **affect** something is to influence it (*a verb*)

The pollution will **affect** the dissolved oxygen concentration.

The pollution has **affected** the dissolved oxygen concentration.

The **effect** of something is the result or consequence of it (*a noun*).

The pollution will have an **effect** on the dissolved oxygen concentration.

Wrong	*Right*
Natural events such as heavy rain may effect the traffic flow. **In some systems, the fish were unable to be eaten, therefore effecting the food chain.**	**Natural events such as heavy rain may affect the traffic flow.** **In some systems, the fish were unable to be eaten, therefore *affecting* the food chain.**
This report examines the affects of natural hazards on communities.	**This report examines the *effects* of natural hazards on communities.**

Putting It Right

- *Effect:*
  Use *effect* or *effects* as nouns. This means:
  ***a, an, the* effect or *the, some, several, a few, many, a couple of* effects.**
  ***Wrong:* This will effect the stream, this has effected the stream.**
- *Affect:* Use *affects* as a verb.
  ***Correct:* This affects something, this will affect something, this has affected something.**
  ***Wrong:* the affects of ...**
- **If it ends in *-ed***, you will almost certainly need **affected, *not* effected.**

Compliment and complimentary/complement and complementary

Compliments/complimentary – where you want to imply flattery.

He complimented the guest speaker on her presentation.
You will receive a complimentary ticket to the dinner.

Complements/complementary

1. The finishing touches to a thing, fitting together, completing
   The formal garden complements the exterior of this superb house.
2. The scientific or mathematical meanings
   complementary angles
   complementary colour
   complementary relationship
   complementary function

Imminent/Eminent

Imminent: soon, impending

Their arrival is imminent.

Eminent: important, distinguished

She is an eminent scientist.

Composed/Comprises

There are two expressions that commonly get mixed up: **is composed of** and **comprises.**

Correct:

It is composed of three parts.
or
It comprises three parts.

Incorrect:

It is comprised of three parts.

Wrong	*Right*
The material used in dental fillings is an amalgam comprised of mercury and silver.	**The material used in dental fillings is made of an amalgam comprising mercury and silver.** *or* **The material used in dental fillings is made of an amalgam composed of mercury and silver.**

> **Putting It Right**
>
> *Comprised*, *comprises* or *comprising* can never be followed by *of*.

Lead/Led

This has become confused because *lead* is pronounced in two different ways:

- The element Lead (Pb)
- Will you lead the team?

This word doesn't follow the same system as *read*, which is what often confuses people.

The most common misuse is in the following:

Wrong	*Right*
This lead to pollution of the stream.	**This led to pollution of the stream.**
This has lead to more interest being shown in the hot air engine.	**This has led to more interest being shown in the hot air engine.**

Putting It Right

Whenever you write *lead*, say it to yourself.

- Does it sound like *led*? If so, write **led** (unless you mean the element Lead, Pb).
- Does it sound like *leed*? If so, write **lead**.

Lose/Loose

In their most usual senses in science writing:

- **Lose** means to *cease to possess* or *misplace*
- **Loose** means *not restrained*

A loose fit.
The cover was loose.

Wrong	*Right*
The breeding pairs will loose their chicks if conditions do not improve.	**The breeding pairs will lose their chicks if conditions do not improve.**

Passed/Past

Examples of use:

Passed

The law that has just been passed states that …
Somatic injury is not passed on to the next generation.

Past

In the past …
Past practices …
Over the past year …
The road runs past the waterfall.

Principal/Principle

Principal: The most important, the highest in rank, the foremost:

The study was made up of five principal sections.
The principal of the institution said that…

Principle: A fundamental basis of something:

Archimedes' principle ...
The chief investigator has no principles.
The principles of the investigation were ...

Their/There

Their is never followed by **is, was, will, can, should, would, could, may, might**.
Wrong: Their was, their is, their could be/should be/would be, their will, etc.
Right: There was, there is, there could be/should be/would be, there will; etc.
On all other occasions (except when you are saying something is 'over **there**'), you are likely to need **their**.

Wrong	*Right*
Their are a number of strategies that countries can take.	***There* are a number of strategies that countries can take (You can't have *their* and *are* together.)**
In there advanced form, they are superior to petrol engines.	**In *their* advanced form, they are superior to petrol engines.**
Under high winds, small boats break from there moorings.	**Under high winds, small boats break from *their* moorings.**

Jargon phrases to avoid in formal writing

Some of the phrases to avoid:

a window of opportunity	**in the long run**
all things being equal	**in the matter of**
as a last resort	**it stands to reason**
as a matter of fact	**last but not least**
at the end of the day	**level playing field**
at this point in time	**many and diverse**
comparing apples with apples	**needless to say**
conspicuous by its absence	**on the right track**
easier said than done	**par for the course**
effective and efficient	**slowly but surely**
if and when	**the bottom line**
in the foreseeable future	**going forward**

Write to inform, not to impress

Guideline: **Write as you would speak in comfortable, serious conversation**.

When they write, many people tend to choose long words, thinking that they are more impressive than the shorter ones used in conversation. The result is pompous and tedious to read.

When writing, think of the clearest way of expressing something. The reader will be impressed not by long words but by clarity.

The following list contains pairs of words that mean the same thing. Technical and professional writers will almost invariably choose the word in the first column, and end up sounding pompous. Your writing will be more direct if you choose the shorter word. But don't avoid the longer words altogether. Avoid using them exclusively, and aim for a mixture of long and short. This will help your reader not to be bored.

Pompous Word	*Short Word*
Anticipate	Expect
Assist	Help
Commence	Start
Desire	Want
Endeavour	Try
Indicate, reveal	Show
Locate	Find
Purchase	Buy
Request	Ask
Require	Need
Terminate	End
Utilise	Use

The split infinitive

Split infinitives aren't nearly as important as they are often made out to be. There isn't, as many people suppose, a rule that says an infinitive should not be split; it is merely an invention observed by them in the mistaken belief that they are showing their knowledge of 'good' writing. Rigorously sticking to such outmoded ideas does just the opposite; it interferes with your ability to communicate effectively. Only misinformed pedants criticise a piece of writing because it contains a split infinitive.

What is an infinitive? This is a verb form. When *to* is followed by a verb (the 'doing' word), it is said to be an infinitive: to differentiate, to prepare, to analyse, etc.

A split infinitive is when words come between *to* and the verb:

They were urged to *seriously* reconsider their stand.

Pedants would insist that this sentence be rewritten. However, the three rewrites sound awkward:

They were urged to reconsider seriously their stand.
They were urged seriously to reconsider their stand (*this is ambiguous).*
They were urged to reconsider their stand seriously.

However, a lengthy interruption is not good:

The political will is lacking to resolutely, wholeheartedly and confidently reform the tax system.

Sometimes, a split infinitive is needed to avoid ambiguity:

He would like to really learn the language.

The alternatives are ambiguous: *He would like really to learn the language* could mean the same as *He would really like to learn the language.*

> **Putting It Right**
>
> **Write whatever sounds the least awkward. Only misinformed people worry about split infinitives.**
>
> However, you need to be aware that some assessors fall into this category. They will hunt split infinitives down and delight in pointing out each one. So avoid writing them if possible.

How long should a sentence be?

Short sentences are more digestible. You can also get into less trouble with their construction. Modern writing and word processor grammar checkers tend to describe sentences of over 25 words as too long. However, don't treat this as an absolute. Your readers will be bored if you deal them equal-length sentences one after another. The occasional longer sentence, if it is well constructed and not overloaded with ideas, will make your writing more interesting.

Variety is important in sentence length. Aim for an *average* of 20–25 words per sentence, but oscillate around the mean.

How long should a paragraph be?

As with sentences, varying the length of paragraphs is another way of avoiding boring your reader. Avoid very long paragraphs; black, uninterrupted text is discouraging.

Many people find effective paragraphing tricky; knowing that long paragraphs are bad, many seem to decide quite arbitrarily on where paragraph breaks should be placed. The result is an incoherent text.

It is difficult to give guidelines how to paragraph effectively, but as a general rule, It can be said:

- One main idea per sentence.
- One theme per paragraph. If there is a natural break in what you are writing, start a new one.
- The first sentence of a paragraph – the *topic sentence* – should introduce the theme of each paragraph.

Verbs and Vivid Language

Vivid language is not something that most people associate with technical writing. Yet if readers are given dull, impersonal prose, they get bored. A lot of technical writing is dull, and much of the problem is to do with the way we use verbs.

Active versus passive voice

Many word processor grammar checkers and writing handbooks tell us to use the active voice of the verb, not the passive. This is not useful advice because most scientists and engineers have no idea what the active and the passive voices of the verb are.

We will now ask the following:

- What is meant by the active and passive voices?
- Is using the passive voice bad?
- What happens when we distort the passive voice and make *really* pompous sentences?

Recognising the active and passive

Many people recognise a verb as the 'doing' word of a sentence. The following sentence has a subject (or actor), *acid etching*; a verb, *removed*; and an object or receiver, *rust*.

Acid etching removed the rust.			Active voice of the verb
Actor	verb	receiver	

This sentence is in the **active voice** because the order of the flow is *actor, verb, receiver*.

If this sentence is turned around, we have the following:

The rust was removed by acid etching.			Passive voice of the verb
Receiver	verb	actor	

When the order of flow is *receiver, verb, actor*, the sentence is in the **passive voice.**

What happens when an active sentence is turned around into the passive voice?

- The emphasis has changed. In the active sentence, the emphasis was on *acid etching*; in the passive form, *rust* is emphasised.
- The order of the flow is reversed.
- The number of words in the verb increases – *removed* becomes *was removed* – as a result of adding forms of the verb *to be*.
- An extra word is needed (*by*).

Is using the passive voice bad?

No. The passive voice is not intrinsically bad, in spite of what grammar checkers and many writing textbooks tell us. We need the passive; it stops us from having to use *I* and *we*. In technical writing we would write, quite naturally:

The *p*H was maintained at 6.8 (*passive*) implying **The *p*H was maintained *by me* at 6.8.**

The active version is unacceptable: **I maintained the *p*H at 6.8.**

But sometimes we can actively choose which voice of the verb to use. For instance, if we were writing a paragraph about bees and their relationship with pollen, we would write:

Bees carry pollen *(active).*

If the paragraph were about pollen, we'd write:

Pollen is carried by bees *(passive).*

Each of these is completely acceptable; it depends on which emphasis we need.

Taking the passive voice one step further: the distorted passive

What *is* bad is to take the passive one step further into a distorted form. Then the verb becomes hidden in a sort of a noun. This happens often in science and technological writing.

Let's consider the progression in these sentences:

Acid etching removed the rust. Turn this around and it becomes:	Active voice	Acceptable
The rust was removed by acid etching. If the verb *was removed* becomes hidden in a sort of noun, it becomes:	Passive voice	Acceptable
Removal of the rust was ...... by acid etching. Hidden verb a missing verb	Distorted passive	Tedious, pompous

Ask someone to insert the missing verb, and the suggestions are always the same: The favourites are **achieved, accomplished, carried out, performed, undertaken.**

Now we've lost the skeleton of the sentence. We've gone from *Acid etching removed*, or *The rust was removed* – both of which are good – to *Removal was carried out, Removal was achieved*, etc., which sound pompous.

This distortion is a common way of writing tedious, impenetrable prose in science. It often sounds completely normal because we've become used to seeing it.

The ohmmeter measured the resistance.	Active voice	Acceptable
Resistance was measured by the ohmmeter.	Passive voice	Acceptable
Measurement of the resistance was carried out by the ohmmeter.	Distorted passive	Tedious, pompous. Not needed.

I measured the leaf area daily.	Active voice	Not generally acceptable
The leaf area was measured daily.	Passive voice	The acceptable style for technical writing
Daily measurements of leaf area were carried out.	Distorted passive	Unnecessary distortion

A method of seeing its absurdity: We are so used to seeing the distorted passive in professional writing that the absurdity of the construction is only obvious when it's seen in an everyday context:

Cinderella dropped the glass slipper.	Active	Acceptable
The glass slipper was dropped by Cinderella.	Passive	Acceptable
Dropping of the glass slipper was carried out by Cinderella.	Distorted passive	Absurd

Putting it right

How do you rewrite the distorted passive?

- If you find yourself using **achieved, accomplished, carried out, performed, undertaken**, you are very likely to be in the distorted passive. So keep these words in mind as danger signals.
- Find the hidden verb. It will be earlier in the sentence, probably be in a word ending with *…ing*, *…tion* or *…ment*.
- Use it to rewrite the sentence, using either a simple passive construction or the active.

Lifeless verbs

Lifeless verbs halt the movement of a sentence. The worst offenders are *exist, occur*, and various forms of the verb *to be.*

Original Lifeless Version	*Rewritten Version*
Increasing temperature occurred.	**The temperature increased.**
The purpose of this report is to describe the different stages of wastewater treatment.	**This report describes the different stages of wastewater treatment.**

Excessive use of nouns instead of verbs

Some lifeless verbs can mutate into nouns, and the pace slows down:

Indicates *becomes* is an indication of.
Suppose *becomes* make the supposition.

Original Lifeless Version with Verb Mutated to Noun	*Rewritten Using Verb*
The colour of the outfall was an indication of severe pollution. **We may therefore make the supposition that ...**	**The colour of the outfall indicated (or *showed*) severe pollution.** **We may therefore suppose that ...**

Subject/Verb agreement

Make sure that the subject (actor) of your sentence agrees with the verb.

Original Incorrect Version	*Corrected Version*
Mazda were the only company that had persevered with the rotary engine concept.	**Mazda *was* the only company that had persevered with the rotary engine concept.**
The greatest loss of lives as a result of a volcanic eruption have occurred through pyroclastic flows and tsunamis.	**The greatest loss of lives as a result of a volcanic eruption *has* occurred through pyroclastic flows and tsunamis.**

Note: The verb is referring to *loss (singular)* not *lives (plural).*

The correct form of the verb

(For simple examples of forms of the verb, see Appendix 2: *The Parts of Speech and Verb Forms.*)

Decisions about the proper use of tense can be confusing. There are no absolute guidelines. Use the following information to decide whether you need the present or the past tense, then use your instinct together with Appendix 1, page 261, to decide on the specific form. Here are suggestions for deciding which tense to use in technical documentation:

1. A form of the ***Past tense*** for describing:
   - ***Procedures and techniques:***
     The samples were fixed in osmium tetroxide (*you are describing a procedure*).
   - **Results** (yours and other people's):
     Brown (2010) found that the numbers of protozoa increased in mature biofilm (*you are describing other people's results*).
2. A form of the ***Present tense*** for describing:
   - ***Established knowledge and existing situations:***
     It has long been known that plants flower (*present*) under environmental conditions that maximise seed set and development.
   - ***For your answers to the research question:***
     The results of this study suggest that *Nitrosomonas* species are (*present*) slow-growing and very sensitive to environmental change.
   - ***Illustrations:***
     Figure 10 shows the effect of temperature on the solubility of the salt.
   - ***Morphological, geological and geographical features:***
     The eucervical sclerites are connected to the postcervical sclerites, each of which is differentiated into a relatively hard sclerotised base and an elastic distal part.

All three paleosols show a greater degree of development than the surface soils. Better development is displayed in terms of greater clay accumulation, higher structural grade, harder consistency and thicker profiles.

- **The theoretical background** (*you are describing established knowledge*)

3. **Specific uses:**
   - The *conditional, subjunctive* or the *imperative* forms can be used when giving recommendations.
   - The *imperative* form is used in procedures or sets of instructions.
   - The *future* form will be needed in the *Materials and Methods* section of a research proposal.

For examples of text, see the following:

- Chapter 6: *A Journal Paper*, the various sections
- Chapter 12: *Procedures or a Set of Instructions*
- Chapter 5: *A Research Proposal*

Putting it right: verb tense

As a general indication, use past tense for almost everything, but use the present for describing the following:

- Established knowledge and theory
- Illustrations
- Morphological, geological and geographical features

Recognising and Correcting Incomplete Sentences

The use of incomplete sentences – sentence fragments – is a common mistake and can greatly irritate assessors.

It's difficult to define incomplete sentences and their rewriting without resorting to classical grammar, but we'll try.

Rule-of-thumb: **Anything between two full stops (that is, a sentence) should sound complete in itself.**

Finite verbs make a sentence sound complete

Most people can find the verb, the 'doing' word, in a sentence. If the main verb in a sentence is what is termed 'finite', then the sentence sounds complete. Rather than define what makes a verb finite, let's just say that if a sentence feels complete, then the verb is finite:

Many factors *affected* the resident population.
The emissions *increased*.
Chapter 12 *presents* the conclusions.

Now consider the following examples. In each case, the first sentence is complete; the second 'sentence' is incomplete because the verb isn't finite. To test this, try saying the second part of each example completely in isolation; it will feel unfinished.

More intensive agriculture will cause an increase in global warming. The reason being that ruminants and paddy fields produce methane.
When running on petrol, the carbon dioxide emissions were higher. Which was an indication of improved mixing and less cylinder-to-cylinder variation.

Recognising them and correcting them

There are two main ways in which incomplete sentences creep into technical writing:

1. **Where the incomplete 'sentence' has a word ending in *-ing* at or near the start of it (*The reason being* is a favourite), then:**
   Method 1: Join it up with the previous sentence with a comma.
   or
   Method 2: Rewrite it using a finite verb. It should sound complete in itself. Just use instinct – it usually works.

Original Incorrect Version	*Corrected Versions* *Corrected Using:* • *Method 1 (a comma)* • *Method 2 (using a finite verb in the second sentence)*
More intensive agriculture will cause an increase in global warming. The reason being that ruminants and paddy fields produce methane.	*Method 1* **Increasing agriculture will cause an increase in global warming, the reason being that ruminants and paddy fields produce methane.** *Method 2* **Increasing agriculture will cause an increase in global warming. The reason is that ruminants and paddy fields produce methane.**
There are a number of strategies that countries can take. For example, promoting non-wood fuel sources, paper recycling and pricing forest products more efficiently.	*Method 1* **There are a number of strategies that countries can take, for example, promoting non-wood fuel sources, paper and pricing forest products more efficiently.** *Method 2* **There are a number of strategies that countries can take. For example, they can promote non-wood fuel sources, recycle paper and price forest products more efficiently.**

2. Where the second, incomplete 'sentence' starts with *Which* (when it's not a question) It's very common to see this in commercial material, for example:
   All of these plans have been designed with you in mind. Which is why you'll find one that's just right for you.

However, this is unacceptable in technical writing.

Method 1: Use a comma instead of a full stop.

Method 2: If inserting a comma makes the sentence too long, rewrite the second part. You can generally start the second sentence with **This is/was/will be** ….

Original Incorrect Version	*Corrected Versions* *Corrected Using:* • *Method 1 (a comma)* • *Method 2 (starting another sentence using* **This is/was/will be** …)
When running on petrol, the carbon dioxide emissions were higher. Which was an indication of improved mixing and less cylinder-to-cylinder variation.	*Method 1* **When running on petrol, the carbon dioxide emissions were higher, which was an indication of improved mixing and less cylinder-to-cylinder variation.** *Method 2* **When running on petrol, the carbon dioxide emissions were higher. This was an indication of improved mixing and less cylinder-to-cylinder variation.**

Summary: incomplete sentences

Methods for recognising incomplete sentences:

- Look for words ending in ***-ing*** at or near the start of a sentence.
- Look for ***Which*** at the start of a sentence, when it's not a question.

Say it out loud. Use instinct. A sentence fragment will generally sound odd. If you say 'This being devastating for the farmers' completely in isolation, it feels incomplete. However, 'This was devastating for the farmers' sounds complete.

- If it feels incomplete, it probably needs rewriting.

When English Is an Additional Language

Anyone who is learning another language knows that writing is the most difficult task of all.

This chapter contains a number of sections that are useful to a speaker of English as an additional language. The following section gives a few basic guidelines about writing a document in English.

1. **Think in English.**
   You will have more success if you think in English and construct your sentences in English. If you compose in your own language and then translate, you might translate the constructions and idioms of your own language into English.

If your language is a European language, your assessor will probably be able to understand a literal translation.

Example: By a German speaker:
This is known since long to affect development of the embryo.

But for languages that are not related to the European languages, there is a much greater possibility of constructing text that does not convey your meaning.

2. **Write short, clear sentences.**
In a technical document, your aim is to present your material as clearly as possible. When you are working in English, it is better to write short, clear sentences, even if they are too short to be regarded as good style. Longer sentences can become complicated and your meaning may not be clear.
3. **Typical mistakes**
It is impossible to give meaningful examples here; mistakes in English will depend on the structure of your first language. But try to build up a list of typical mistakes that a speaker of your language makes; e.g. by listing corrections made by your assessors.

How to write the English of your own specialist field

Actively work at improving your written skills. If you do not understand English well, or you can only speak it slightly, English can become a background noise. When this happens, your written skills will stay fixed at a low level. It is very easy to become involved in your experimental work, and speak mostly in your own language to friends and family. You need to actively work at increasing your skills. Here are some suggestions about how to do it:

1. **Read as many papers in your field as you can.** You will find that there are sentences and constructions that occur so commonly that you can modify them and use them for your own writing.
2. **Look up words you don't know in a dictionary.** It is too easy to get only the general sense of a paper when reading it, rather than using it to increase your word power. For technical terms, you may need to use a scientific or technical dictionary or encyclopaedia.
3. **Write things down.** Words, constructions and idioms are easily forgotten if you don't actively work at remembering them. Making a list of them does two things:
   - Writing them down makes you remember them.
   - You have a reference list of useful material.

   Later, when you become more skilled in English, this list will look unbelievably simple; but at this early stage, it is essential.
4. **Actively listen to seminars and lectures.** Keep adding to your list while you listen. This will also give you the sort of spoken English that is used in formal presentations. This can be more idiomatic than journal paper English and can cause more problems because you can't go back and read it again.

Section 5

Presenting Your Work Orally

19 A Seminar or Conference Presentation

This chapter aims to help you develop the competencies needed to deliver a presentation at a seminar or at a conference. It covers the following:

- The constraining factors of a presentation
- Features that audience members most dislike in presentations
- Strategies for beginners
- The basic activities: time sequence
- Structuring your presentation
- Designing your slides (PowerPoint or other presentation software)
- Delivering your presentation
- Running out of time
- Answering questions
- Common mistakes
- Checklist

It can be nerve-wracking being faced – perhaps for the first time – with having to present your work orally to an expert audience. The problem can be even worse if English is an additional language for you.

The most common fears are of (1) looking foolish, (2) forgetting what you need to say and (3) feeling exposed because of the primitive fear of everyone's eyes are looking at you. Advice on public presentation of scientific and technical material often appears to forget that much of it can't be put into practice by a beginner. Many students are concerned only about surviving the experience with their credibility intact.

The objective of this chapter

1. This chapter accepts that the thought of public speaking is unsettling or even terrifying. It gives guidelines from this standpoint and tries to avoid giving advice that is of no use to a beginner.
2. The chapter also gives guidelines on common mistakes to avoid. Being aware of the things that people can do when they are nervous can help you avoid or minimise them. This material is based on the feedback that I've given to the graduate student presenters of the many hundreds of presentations that I've seen over the past 15 years.

What this chapter does not do

While it gives guidelines for effective presentation graphics, it doesn't suggest pictorial designs. There is a wealth of online information available; be careful, some is good, some less so.

Writing for Science and Engineering.
DOI:http://dx.doi.org/10.1016/B978-0-08-098285-4.00019-4

The Constraining Factors of a Presentation

1. At a conference, to be able to accommodate the feeling that you're on trial in a strange room with possibly strange equipment.
2. To present your work orally to an audience that might be expert, non-expert or mixed.
3. To be able to present within a fixed time limit. This will probably seem beforehand to be very long but is usually too short when you're doing it.
4. **To remember about a conference punctuation:** A significant number of people in the audience may be used to *reading* English but not be very good at *hearing* and understanding it (see *Eighteen Design Principles for Your Slides*, page 239).

Features that audience members most dislike in presentations

These are common faults. They're not listed here to make you nervous: it is useful to be aware of them so that you can avoid some or all of them, if possible.

1. Too much text on the slides (always the most greatly disliked feature).
2. Text is too small to be read.
3. Text is made up of full sentences instead of brief points.
4. Illustrations are too small, too complicated or not labelled informatively.
5. Material is not presented logically.
6. Poor background and/or colour choices: interference with the text and illustrations.
7. Inappropriate animation effects, e.g. zooming bullet points, unneeded transitions between slides.
8. Reading too much: from either notes, the monitor or the screen.
9. Speaking too quickly and/or too quietly.
10. Standing so that part of the audience's view of the screen is blocked.
11. Not looking at the audience, looking at only one part of the audience or looking at only one person.
12. Continuously swooping the laser pointer over the screen.

Strategies for Beginners

Overview

1. The most effective strategy: Know the first minute of the presentation by heart and the rest of it very well; a sip of water or deep breathing may help.
2. Don't use a written script.
3. Don't worry if your voice sounds funny; it probably won't sound that way to the audience.
4. Take every opportunity to give talks; don't try to avoid them.
5. Be aware of the mistakes people make when nervous.

1. **What can you do about being nervous?**
   Accept that you are probably going to be nervous and have strategies to deal with it. In spite of what you may read in some books that say you can convince yourself you're not nervous, there don't seem to be any foolproof methods of doing this.
   The most effective remedy is to accept that if you don't have much experience, you are probably going to be nervous, and that it might hit you very unexpectedly. **Have these strategies in mind to deal with it:**
   - **The most effective strategy** is to know the first minute of your talk by heart, i.e. while your brain is on autopilot, and know the rest of your talk very well.
   - **Knowing that you have a well-prepared presentation.**
   - **Some other techniques that might help** include having a sip of water just before speaking and using deep breathing exercises (it helps some people but beware of hyperventilating).
   - **Preparing for unexpectedly being hit by nervousness**. It also helps to assume that, even if you think you're not going to be nervous, you might suddenly be hit with it at the very last minute, often when the chair announces you.
2. **Don't use a written script.**
   Don't fall into the trap of taking a full script to the presentation, thinking you might forget what to say. Because of nervousness, people often end up reading solidly without looking up. It looks very unprofessional and, if you are being graded for your presentation, you'll be marked down.
   The best way of avoiding a written script is to make sure that everything you need to remember is on your slides, either as very brief points or as illustrations. Then you can get your prompts by glancing briefly at either the monitor or the screen.
3. **Don't worry that your voice might sound funny.**
   You may feel your voice might crack embarrassingly. Don't worry. People won't notice. It is very common to find that members of the audience will tell you afterwards that it was unnoticeable.
   Remember, your voice always sounds very different to you than it does to another person because much of it is transmitted to you through the bones and cavities in your head. What sounds to you like a funny voice may not sound very different from your normal voice to an audience.
4. **As a long-term strategy, present often.**
   Take every opportunity to give talks. It's a tough strategy, but it's only by frequent practice that you will eventually become less nervous. Even very good presenters have felt nervous at the beginning of their careers.
5. **Be aware of the strange things that people can do when they are nervous**, and try to avoid doing them.

The Time Sequence of Activities in a Science Presentation

Overview

Here is a simple list of the sequence of activities in a presentation:

1. **Greet the audience**, say who you are and what you are going to talk about (see *Choice of Words*, page 243).
2. If needed, provide an **Overview/Summary slide** of the whole talk.

3. **Introduction:** Should contain *very brief outlines* of the **context**, the **motivation** and **objective** of your work:
    - **Context:** Overall view of the field.
    - **Motivation:** *Why* you are doing this work: the gap in the knowledge that your work will fill.
    - **Objective**: *What* you are aiming to achieve, the purpose of your work.
4. *Probably:* your **method of approach**.
5. What your **results** were.
6. How you **interpret** your results.
7. The final few minutes, **give your conclusions** from your work in a brief list.
8. **Thank your audience** in very few words (see *Choice of Words*, page 243).
9. **Stop. Remain standing. Don't ask for questions if it's at a conference.**

Structuring Your Presentation

Overview

1. **The major fault with many presentations is the structure of the material,** not the slides or the manner of delivery.
2. **Plan the following:** Your main points, illustrations, the logical thread through it, getting your prompts, the amount of detail, how to greet the audience and take your leave, the structure and the overall sequence of activities.

A presentation will be assessed on whether the audience can readily access and understand the main information. Without a good structure, no amount of colourful slides and confident speaking will make for a professional presentation.

Common major mistake: no obvious logical structure.

Unstructured facts, the main points not being obvious and no clear framework or story are often the main reasons for an ineffective presentation, not the slides or the way you deliver it.

Planning the structure of the presentation

Plan these features:

1. **Start by deciding the main points of your talk.** Be *very* selective.

Common mistake: Trying to present too much information causing you to almost certainly run out of time and probably confuse the audience.

2. **The illustrations you're going to include:** Schematics, photographs, graphs, etc. Plan the storyline of your presentation around them, as when planning a journal paper (see *How to Start Writing a Journal Paper*, Chapter 6: *A Journal Paper*, page 84).

> **Common mistake:** Far too much text on the slides. Scientific and technical audiences do not like slides that are crammed with text (see *Features that Audience Members Most Dislike in Presentations*, this chapter, page 232). They assess complex material much better through illustrations.

3. **A very logical structure, with a clear storyline.** Choose the illustrations so that the logical thread is obvious.

> **Common mistake:** No obvious framework or storyline is running through it.

4. **The design of your slides.** Choose the template and the colour combinations very carefully (see page 240). For each slide, plan:
   - The illustrations
   - An informative heading
   - Whether very brief written explanations are needed
   - The major points that you need to make while you are speaking about each one.

   See *Designing Your Slides (PowerPoint or Other Presentation Software)*, page 238.

> **Common mistake:** Poorly designed slides that interfere with the audience's understanding of the material.

5. **Getting your prompts: how to make sure you remember everything you need to say without using notes.** Design your slides so that by briefly glancing at the computer monitor or the projection screen, you'll remember all of the points you need to cover.
   Try not to use a complete script of written notes. Many people do this as a form of security and intend not to use it, but because of nervousness, they end up reading continuously.

> **Common mistake:** Reading solidly without looking up from a script, the monitor or the screen.

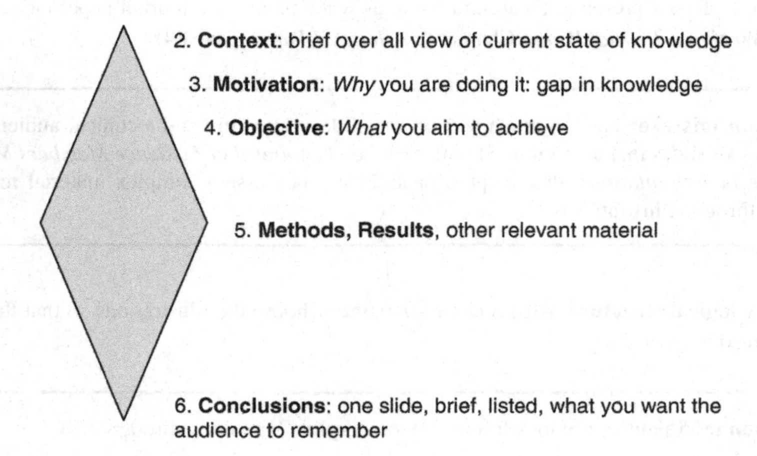
1. Titleslide
2. Context: brief over all view of current state of knowledge
3. Motivation: Why you are doing it: gap in knowledge
4. Objective: What you aim to achieve
5. Methods, Results, other relevant material
6. Conclusions: one slide, brief, listed, what you want the audience to remember

Figure 19.1 Structure your talk like a diamond: present material in this order (1–6).

6. **The amount of detail**
   If you have also written a document on the same topic, be aware that the presentation will need far less detail. You'll need to be very selective. It is a far greater achievement to be able to express a complex idea clearly than to drown the audience in detail.

> **Common mistake:** Far more detail than is needed; the audience becomes lost, and you run out of time.

7. **Professionally greeting the audience at the beginning and taking your leave at the end.**
   See *Delivering Your Presentation*, page 242.

8. **Structure your presentation as a diamond** with the concise material at the beginning and at the end. Present material in the order shown in Figure 19.1.

9. **Overall sequence of activities and information:**
   (a) **Title slide and greeting**
   Make sure your title slide shows your name, title of the presentation, your institution and city. Include your country if it's an international conference.
   Look directly at the audience (not at the title slide), and then say who you are and what you are going to talk about.
   See *Delivering Your Presentation, Opening and Closing*, page 243.
   (b) **Outline or Overview slide: a brief summary of all the material**
   Some people like these, some don't. It's your choice.
   If you do decide to include an outline slide, make sure it's informative. It should give a summary in a few words of each aspect of your presentation.
   Briefly list the main points that you are going to cover, and then *briefly* explain them.
   Make sure that the information you give on your list of topics is real information. Avoid meaningless headings (Figure 19.2).

Effective list of overview points	Ineffective list of meaningless overview points
Failure of Impacted Sandwich Composite Aircraft Panels **OVERVIEW** **Objective** To determine the size of damage tolerated, without failure stress of the panel being reduced below allowable design limit. **Material tested** Identical to that used on trailing-edge wing flaps of Boeing 747-400 **Methods - comparison of:** 1. Analytical model to determine wrinkling stress 2. Experimental testing: panels impacted using pneumatic gun and solid indentor **Results and Conclusions** 1. Critical damage size: 22 mm. 2. General empirical model derived to approximate wrinkling stress behaviour.	**Failure of Impacted Sandwich Composite Aircraft Panels** **OVERVIEW** **Introduction** **Methods** **Results** **Conclusions**

Figure 19.2 The contrast between an effective and an ineffective overview slide.

Common mistake:

To show only a list of bare headings (for example, *Introduction, Methods, Results, Conclusions*) while you say '*This shows an outline of my talk. First I'm going to give an introduction, then I'll describe the methods, then give you my results and finally my conclusions.*' This is a statement of the obvious to the audience and a waste of time for you.

(c) **Introduction**: It should contain the **context** (a brief overall view of the field) followed by the **motivation** of your work (*why* you are investigating it; that is the gap in the knowledge that your work will fill).

(d) Consider having a separate slide titled **Objective** *or* **Aim** giving a clear, brief statement. This is very useful to the audience.

Common mistakes:

- Spending far too long on the introductory material and running out of time. Most of the talk should concentrate on what you have found out. Give only enough background for the audience to be able to place your material in context.
- Having no clear objective.
- Little or no logical structure; no framework on which to hang the story.
- Without a *Conclusions* slide, you leave the audience with no clear main messages to take away.
- If you run out of time, rushing through your remaining slides because you have no *Conclusions* slide (see page 246).

(e) **After the objective, present the main body of the information.** Make sure that in each section, you present the material in an obvious framework, with a clear storyline running through it.

(f) **A concluding slide titled *Conclusions*.** A concise list of the conclusions you draw from your material: the points you most want the audience to remember.
This is a very effective way of closing your talk; moreover, this slide will be very important for finishing in a hurry if you run out of time (see page 246).

Designing Your Slides (PowerPoint or Other Presentation Software)

Overview

This section gives 18 fundamental design principles for your slides:

1. How many slides?
2. Make slides uncrowded.
3. Not too much text.
4. Any text should be brief points.
5. Font size large enough for audience to read easily.
6. All key points must be on your slides, not just spoken.
7. Make the key points reinforce what you are saying at any one time.
8. Illustrations clear enough to be readily interpretable by audience.
9. Graphs may need a brief written explanation of meaning.
10. Simple background or template.
11. Maximum contrast with background.
12. Choice of colours.
13. Shaded backgrounds.
14. Use animation functions intelligently.
15. Speaking the detail while showing only general points on screen.
16. Using citations.
17. Proof read the slides.
18. Show where you are in the scheme of the talk.

Eighteen design principles for your slides

There is a wealth of online material about slide design; some material is good and some is questionable. Here are 18 fundamental principles to consider:

1. How many slides?

Use as many as you think appropriate to the subject matter and length of the presentation. Some sources say one slide per 2 min. This is misleading because more can be used provided they are used effectively. However, underestimate the number rather than assume that you can use a lot.

Common mistakes:

Far too many slides, therefore running out of time; hurrying through the remaining slides. If too few slides, the material looks very thin.

2. **Slides should not be overcrowded and cluttered with information.**

Common mistake: *Far* too much on one slide.

3. **Do not use too much text.**
Science and technical audiences respond far better to illustrations than to reading a mass of text.

Common mistake: Too much text, full sentences and not enough schematics and illustrations.

4. **Any text should be in the form of brief points, not full sentences.**

Common mistake: Long, complete sentences.

5. **The font size should be large enough for the audience to be able to read it easily.**

Common mistake: Font is far too small for the audience to read. Font should be at least 25 point, even for a small seminar room. This will look unnaturally large on your monitor while you are preparing the presentation but will be of minimal size for the audience. **PowerPoint** defaults are 32 point for the main text and 44 point for the title.

6. **All your key points must be visible in brief form on your slides, not just spoken.** It is helpful to all members of the audience but is also essential for two reasons: (1) you can readily get your prompts by glancing at the screen or monitor; (2) it is helpful for the audience members who may not be able to readily understand spoken English.

Common mistake: Forgetting important material that you intended to include because of a lack of prompts.

7. **Make sure that at any point in the talk, you reinforce what you are saying by showing the relevant text and illustrations on the screen.**

> **Common mistake:** Speaking a lot of detail with nothing on the screen to reinforce it. Many members of an audience, particularly at an international conference, may not be able to readily understand spoken English.

8. **The audience must be able to see the relevant detail of the illustrations**: clear points, lines, axes and labelling.

> **Common mistake:** Illustrations that are too complex, too small, too finely drawn or over-enlarged for their resolution. Don't import them straight from a document; they will need thicker lines and larger labelling. Diagrams from the web often have too low of a resolution.

9. **Graphs may need a brief written explanation of the main point(s).** This will be helpful to all members of the audience, particularly those who may not understand spoken English very well such as those at an international conference. They will appreciate a text box under each graph that *briefly* summarises the main conclusion(s) of the graph. An informative heading to the slide also helps.

> **Common mistake:** Only a long, detailed *spoken* explanation of graphs and other diagrams.

10. **Use a simple, uncluttered background or template.**

> **Common mistake:** Complex templates and colours interfering with the readability of both text and illustrations.

11. **Aim for maximum contrast with the background: very dark on very light or vice versa.**

> **Common mistake:** The contrast is often not enough to enable the audience to see the material clearly. What looks clear on a good monitor can look quite different when projected.

12. **Be careful with your choice of colours** for both text and background. Some colour combinations give poor clarity (e.g. red on a blue-hued background).

> **Common mistake:** Material is unclear because of colour interference. Again, be careful: it may look good on your monitor but not when it's projected.

13. **Use shaded backgrounds carefully**. They are very popular and effective, both for the overall background of the slides and for individual text boxes. However, they can lead to problems.

> **Common mistake:** Text and illustrations can easily become unclear on the various areas of shading.

14. **Use the animation functions intelligently**, e.g. to build up complex diagrams or schematics so that each element is brought in sequentially while you speak about each.

> **Common mistake:** Using wild animations purely for dramatic effect. They can be irritating to a technical audience. Make points just appear rather than slide, bounce or zoom in.

15. **Make sure you are not speaking lots of critical detail while showing only general points on the screen.** Every slide should be informative.

> **Common mistake:** Showing general points on the screen and speaking the detail, particularly when describing what a graph shows. Help the audience by having a short line of text under the graph giving the conclusions of the graph.

16. **Use abbreviated citations to other people's work.** If you don't use citations, you are implying that it's your own work. As long as the citations are shown on screen, you can use very small font, e.g. 14 point.

> **Common mistake:** To use an illustration from another source and not cite it. By doing this, you are implying that it is your own work.

17. Proofread your slides.

Common mistake: Typos and spelling mistakes look careless and very unprofessional. You may not notice spelling mistakes, but there will be people who do, some of whom you may need to impress.

18. Show clearly where you are in the scheme of the talk.
Suggestions:

- Use an informative heading for each slide.
- Use a bar at the bottom or side of each slide that highlights where you are at any point in the scheme of the talk. This is invaluable for the audience during a complex talk but pointless if the bar just shows *Introduction, Methods, Results, Discussion.*

  Note: PowerPoint does not have an automatic function for doing this. You'll need to make a text box, insert one on each slide, and manually highlight the specific words depending on the individual slide.
  For example, for a presentation on electric vehicles:

Introduction History **Plug-ins** Hybrids Other types Batteries Future Conclusions

Common mistake: People in the audience cannot perceive your progression through the material.

Delivering Your Presentation

Overview

1. Choice of words
   - Open and close professionally.
   - Use spoken English style, not written.
   - Don't be afraid of using *I* or *We.*
   - Use simple, clear words with correct technical vocabulary.
   - Use verbal hints.
   - Don't read out sub-headings.
   - Reinforce main points by showing them on screen.
   - Sound as though you're interested in your work.
2. Your voice
3. The way you stand
4. Your hands
5. Interaction with the screen
6. Looking at the audience
7. Using a pointer
8. Needing to pause

1. *Choice of words*

   a. **Open and close the presentation in a professional manner.**

   Opening:

   Be professional. This doesn't mean being stiff and stilted; it means all of the following points:

   - When you introduce yourself, show your title slide, and try to look straight at the audience, even if it scares you (see *Eye Contact*, page 245).
   - Don't fluster, mutter or giggle.
   - Say: *Good morning/afternoon. I am (your name). I'm going to present my results on.../ talk about (your topic).*
   - Then move smoothly into the next slide.

Common mistakes:

- Forgetting to greet; unprofessional greeting.
- Turning immediately to the screen and reading your name and the title.
- With only the title slide on the screen, giving far too much verbal description of the background and motivation to your study. These details should be presented on the subsequent slides.

Closing:

Put up the Conclusions slide (see page 238). *Briefly* run through the points on the slide. Try to avoid reading it word for word. Then just nod your head and say *Thank you.*

Common mistake: Unprofessional closing such as *Well, that's all I've got to say, really* or *That's it – so-um – thank you.*

   b. **Use spoken English style, not written.**
   You should give the impression that you are speaking to the audience, rather than reciting a written script. On the other hand, don't become too colloquial or matey. You'll lose credibility.

   c. **Don't be afraid of using *I* or *We* in a presentation.**
   It livens up a presentation. But don't overdo it, or it might sound childlike.

   d. **Use simple, clear words but include the correct technical vocabulary.**
   Think in terms of the style of comfortable, serious conversation. Imagine yourself explaining your work across a table to a colleague, comfortably and without using colloquialisms.

   e. **Verbal hints are important.**
   They enliven the talk. *This is important because...; This was an interesting result, because...; This was an unexpected result*, etc.

   f. **Don't read out sub-headings.**
   This shows that you've planned your talk in written terms instead of spoken ones, and that you're reading a script.

Common mistake:

With novice speakers, it is common to hear oddities such as:
Sampling methods. Three sampling methods were used…
Design objectives. The design objectives were….

g. **Reinforce the main points of what you are saying by simultaneously showing it in brief point form on the screen.**
It is not enough for your visual aids to show only diagrams and illustrations. You also need to have text that echoes in brief point form what you are saying at any time.
For example:
- Ineffective: just to say
There were three reasons why we modified the test rig in this way. The first was we found that… etc., meanwhile counting them off using hand-waving body language, with nothing on the screen. Remember that many in the audience may not fully understand what you are saying.
- Instead, prepare a visual aid that lists it in point form, and expand on each one while you speak. Note: It's important not to read the slide woodenly word for word.

Design of Slide	Possible Spoken Material
Reasons for modifying the test rig:	There were three reasons why we modified the test rig.
1. xxx	The first was …
2. xxx	The second reason was …
3. xxx	The third reason was that…

Common mistake: Insufficient planning of your visual material to ensure that your words are reinforced by the material on the screen leads to an ineffective presentation.

h. **Sound as though you are interested in your work.**
The audience will find your presentation and your work more appealing if you can manage to sound interested in it. Just try to convey enthusiasm by making your voice and gestures animated without becoming over-enthusiastic. Use verbal hints, too (see page 243).

2. ***Your voice***
If you don't have a microphone, your voice should (1) be louder and slightly more deliberate than in normal conversation; (2) sound lively and convey your interest in your work and (3) not have any verbal tics, e.g. *basically, you know, sort of, like, uuuum, anduuh.*

Common mistakes: When nervous, your voice can speed up, be quieter than usual; become monotonous, crack or wobble.

Remedy for wobbly voice: Don't worry about it; it has to be very bad before it is noticeable to the audience. If you don't believe this, ask your friends after your talk if they noticed anything. They probably didn't.

If English is an additional language: The problem of a quiet, rapid voice is a common one because of either (1) nervousness about making grammatical mistakes or (2) feeling that you are fluent enough but forgetting that you may have a strong accent or slur words together.

Slow down, and try to speak deliberately, positively and more loudly.

Remember, your audience has come to learn about your research; it doesn't matter about grammatical mistakes as long as they can hear and understand it! It is better to say something ungrammatical in a strong voice – your audience will understand and be sympathetic.

3. ***The way you stand***

Try to stand and move naturally. Be aware of the audience members and try not to block someone's view.

Common mistakes: Standing sideways and not looking at the audience; blocking the view of part of the audience, particularly in a small room; standing rigidly at the computer; crouching over the monitor.

4. ***Your hands***

Beginners often comment that they suddenly become conscious of their hands and don't know what to do with them. You will probably have a pointing device, which takes care of one hand. Make sure you don't make the usual mistakes with the other.

Common mistakes: Putting them on your hips or hooking the thumbs into your waistband or pockets; fiddling with a pen, or with something in your pocket; worse – clicking a pen.

5. ***Interaction with the screen***

It is essential to the audience that you interact with the screen and point out the notable features of your visual aids to the audience. But don't look at the screen for long periods.

Common mistakes: Forgetting to point and leaving the audience to navigate their own way through a complex diagram; looking too much at the screen.

6. ***Looking at the audience (often called eye contact)***

Try to scan around the audience as much as possible, even though it can be very nerve-wracking seeing everyone looking at you.

Useful tip: If you find that looking at people's eyes is difficult, try looking at the chin/neck region instead. No one will notice that you're not looking straight at them unless they are in the very front of the audience.

Don't scan above the heads of the people in the back row. This is sometimes given as advice, but you will look spaced out.

Common mistakes: Hardly looking at the audience at all or only very occasionally; looking at the screen too much; looking at only one part or member of the audience.

7. ***Using a pointer***
   You will probably have a laser pointer or – more rarely – a pointing stick. Beginners often don't like using a laser pointer because shaking hands will be obvious. If it is the only pointer available, which is probably the case at a conference, make sure that you don't circle it continuously on the screen in an attempt to disguise the shakes. Just find the item on the screen, and then keep the pointer firmly in the right place for about only two seconds.

Common mistakes: A continuously swooping laser beam (members of the audience find it very distracting); forgetting to turn off the laser spot when you turn away from the screen so that it wanders around; pointing only with your finger (remember your line of sight is different from the audience's); turning your back on the audience too much.

8. ***Needing to pause***
   If you suddenly forget what you want to say, make sure that the audience doesn't notice that you feel flustered. Just say nothing and control your body language. Remember, a pause that seems very long to you will hardly be noticed by the audience if you don't bring it to their attention.

Running Out of Time

Overview

- Common reasons for running out of time
- Formula for finishing professionally in a hurry: go immediately to the *Conclusions* slide.

People frequently run out of time when giving a presentation and then need to finish hurriedly. The following are the most common reasons for running out of time:

1. **Spending too long on the introductory material.**
   Remember, people in the audience *most* want to hear about your results, not lengthy material about other people's work. Give just enough background for them to understand the context and motivation of your work.
2. **Explaining some slides more than you expected to.**
   Remember, it is very easy for unplanned material to come into your mind in the stress of the moment. Remedy this by planning the points you need to make at each slide, and train yourself not to deviate.

3. Practicing it too fast.
When you practice, speak it out loud. If you read it to yourself or whisper it, you go faster than you would when you speak it.

Formula for finishing in a hurry
If you need to finish quickly, don't panic. Use the following three steps. You'll be able to deliver the main points of your presentation to your audience without hurrying, and you'll look professional:

Step 1: When the final signal sounds, don't look disturbed, flustered or say anything; smoothly finish the sentence you're saying and then go immediately to your *Conclusions* slide.
Step 2: Then say something like: *I'm sorry, I don't have the time to show you all of my material. However, the conclusions from this work were…*, and very briefly run through the conclusions.
Step 3: If time is very short indeed, show the conclusions slide for the audience to read, and close your presentation.

Common mistakes: Drawing attention to yourself by flustering; rushing through the remainder of your slides looking for relevant ones.

How to find your *Conclusions* slide immediately
You can do one of two things if you are using PowerPoint:

1. The best solution: Before your presentation, note the number of the *Conclusions* slide (on your hand if you think you'll forget it), and then when you need to finish, type its number and hit *Enter*. You may think this sounds unlikely to work, but it does. Try it!
2. Copy your *Conclusions* slide so that it appears twice: once in its normal position, and also as the very final slide. Then if you need to finish hurriedly, hit the computer's *End* key. This will take you to the final slide.

Answering Questions

Overview

1. Work out the questions you may be asked. Don't be taken by surprise.
2. Use a supplementary set of slides.
3. Make sure you understand the question correctly.
4. Don't be afraid to ask for further clarification.
5. Repeat the question if you think the audience may not have heard it.
6. Be honest if you don't know the answer.
7. Don't be afraid to admit to research problems, but do it positively.

Many students are nervous at the thought of answering questions at the end of a presentation, for two reasons: the fear of not knowing the answer and the fear of not understanding the question, possibly because they have English as an additional language.

The following points will help:

1. **Work out possible questions beforehand.**
2. **Include a supplementary set of slides at the end of your presentation.**
   These can be more detailed and in a smaller font than those of your main presentation. You may not need them; however, it looks very professional to move immediately to a slide that you can use to answer a question.
3. **Make sure that you understand the question correctly.**
   This can be particularly problematic and worrying for speakers of English as an additional language. Do not answer a question until you are sure you have understood it. Be prepared for the questioner who gives a mini-lecture where the main point can often be buried within a long discourse. Ask for further clarification if you have not understood it.

   Suggested wording, said positively:

   I'm sorry; I didn't understand that. Could you repeat it, please?

4. **Don't be afraid to ask for further clarification.**
   If, after the questioner has tried to clarify it, you still don't understand it, don't panic. Turn to the chairperson and calmly ask him or her to clarify it.

   Suggested wording, again said positively:

   I'm sorry; I still didn't understand that. Could you help me, please?

   Make sure you say this strongly, not pathetically.

   Now it's the chair's task to explain it to you. Don't be embarrassed by it. This is completely normal; one of the functions of a chair is to ensure that the session goes smoothly.
5. **Repeat the question if you think the audience may not have heard it.**
   Say The question was *'How does the...?'*, and then answer it.
6. **If you don't know the answer to a question, be honest.**
   Don't try to fudge your way through an answer and hope that the audience doesn't notice that you are trying to cover up. It is always very obvious when a speaker is doing this.

   Either say in a positive voice that you don't know, or offer to find out the answer. This is a clear indication of honesty and willingness to communicate the research, and that you are confident in your work.

   I don't know the answer to that question, I'm afraid

   or

   I don't have the answer to that at the moment, but I'll find out for you by tomorrow
   or something similar.

7. **Be honest about your research problems but not negative.**

   Don't be afraid to mention – briefly, objectively and without emotion – any difficulties you may have had with your work. It's not a sign of weakness; everyone in the audience will be able to relate to it, and someone may be able to help. But make sure you don't present yourself as self-pitying.

Checklist for a Presentation

Are you a beginner?

- ☐ Do you know the first minute of the talk by heart?
- ☐ Have you avoided using a written script?
- ☐ Do you know the time sequence of a presentation?

Planning and structuring the presentation

- ☐ Have you decided the main points of your talk?
- ☐ Have you decided on your illustrations and planned your talk around them?
- ☐ Have you been really selective and concentrated on your main points?
- ☐ Have you planned your visual aids so that each key point will be simultaneously spoken and displayed in point form on the screen?
- ☐ Is there a very logical structure, with a clear storyline?
- ☐ Have you checked that you will be able to get all your prompts off your slides?
- ☐ Have you avoided too much detail?
- ☐ Have you planned how to professionally greet and take leave of the audience?
- ☐ Will your *Introduction* very briefly cover the context, motivation and objectives of your work? Perhaps a separate slide for the objectives?
- ☐ Have you got a separate slide for your concisely listed *Conclusions*?
- ☐ If you have an *Overview* slide, does it show meaningful information?

Planning

- ☐ Have you been really selective and concentrated on your main points?
- ☐ Have you planned your visual aids so that each key point will be simultaneously spoken and displayed in point form on the screen?
- ☐ Do you have a title overhead or slide, giving your name, title of the presentation, your institution and city? And country, if it's an international conference?
- ☐ Are you aware of what people can do when they are nervous, and have you planned to avoid them?
- ☐ Have you got a final slide briefly listing your conclusions?
- ☐ Are you planning to use notes? Could you possibly do without notes and take your cues from the audiovisual material?

The presentation itself

- ☐ Have you planned how to professionally greet and take leave of the audience?
- ☐ Are you using the style of spoken English, not pompous written English?
- ☐ Have you planned to look up as much as possible?

Design of your slides

- ☐ Have you got too many to fit into the time?
- ☐ Is the font at least 25 point?
- ☐ Have you avoided cramming too much onto the slides?
- ☐ Is the text in the form of brief points? No long sentences?
- ☐ Are any of your slides too crammed with information? Are all of the slides simply designed and uncrowded?
- ☐ Have you checked that the colour and background of your slides is not going to interfere with your material?
- ☐ Have you aimed for maximum contrast with the background?
- ☐ Will the audience be able to see everything on each slide?
- ☐ Are the text and illustrations large enough for the audience to understand?
- ☐ Have you checked that you can get all your prompts off your slides?
- ☐ Does your title slide show your name, title of the presentation, your institution and city? And your country, if it's an international conference?
- ☐ Will your *Introduction* very briefly cover the context, motivation and objectives of your work? Perhaps a separate slide for the objectives?
- ☐ Have you got a separate slide for your concisely listed *Conclusions*?
- ☐ Are all your key points on the slides?
- ☐ Will each key point be simultaneously spoken and backed up in point form on the screen?
- ☐ Under each graph, have you got a brief written explanation of what it shows?
- ☐ Are you using the animation functions intelligently?
- ☐ Are you planning to avoid showing general material on the screen while speaking the detail?
- ☐ Are you using citations for the material that you copied?
- ☐ Have you proofread the slides?
- ☐ If you have an *Overview* slide, does it show meaningful information?

Delivering your presentation

- ☐ Do you think the volume and speed of your voice are sufficient?
- ☐ Will you try to avoid blocking the screen in a small room?
- ☐ Have you planned to look at the audience when appropriate?
- ☐ If you are using a laser pointer, are you planning to use it so that the spot doesn't constantly flicker over the screen?
- ☐ If you lose your place, have you planned not to fluster and calmly find your place again?
- ☐ Do you sound as though you're interested in your work?

Answering questions

- ☐ Have you worked out beforehand the possible questions?
- ☐ Have you got a supplementary set of more detailed slides that you can use for answering questions?
- ☐ Do you know how to deal with not understanding the questions?

20 A Presentation to a Small Group

This chapter covers the following:

- The basic principles for preparation
- A Ph.D. oral examination
- A presentation to a review panel, e.g. a design interview, a presentation to a funding organisation
- Checklist

Occasions When You Might Present to a Small Panel

- A Ph.D. oral examination
- A design interview
- A presentation to your funding organisation

Constraints

The meeting is likely to take place across a table and may have no means of projecting visual aids.

Basic Principles for Preparation

1. Visualise yourself and your material through the audience's eyes.
2. Work out beforehand the questions you may be asked.
3. Identify the main points of the work and its strengths.
4. Identify the key weak points and problems, and prepare yourself for questions about them.
5. Think graphically. Clear, graphical visual aids are an effective means of making points and answering questions.

A Ph.D. Oral Examination

1. Allow adequate time to become thoroughly familiar with the material and thesis again. There is likely to be a gap of months between submission and oral, which is enough time for the detail and structural plan of your thesis to become blurred. It may take more time than you might think to become fully refreshed.

Writing for Science and Engineering.
DOI: http://dx.doi.org/10.1016/B978-0-08-098285-4.00020-0

2. **Prepare a summary of the work that you've done and its significance.**
   Keep in mind what you've done, how you've done it, what's new about your research and what's significant about it. After the initial small talk to make you feel more comfortable, Ph.D. oral examinations often start with a request to the student to summarise his or her work. To prepare for this beforehand, you need to be able to stand back from the minute detail and prepare an overview.
   Possible questions: Make sure that, amongst other things, you can answer the following:
   - *What is the significance of your work?*
   - *What skills did you develop?*
   - *If you were to do it again, would you approach it differently?*
   - *Where do you see it leading?*
3. **Make sure you can navigate your way around your thesis without hesitation.** You may need to refer to it to answer questions.
4. **Near to the date of the examination, do a literature search for any new work that may have come out.** The months between submission and the oral exam can mean that you may be unaware of significant new developments. There is an opportunity to impress here: relate the new work to yours, and decide where you want the discussion to go.
5. **During the examination, if you are asked a question that needs deliberation, allow yourself time to think without getting flustered.** Don't let the pressures of the moment force you into a hasty answer. Your assessors will prefer a period of thought followed by a reasoned answer to an unconsidered, hasty one.

A Presentation to a Review Panel

An example of this type of presentation is for an engineering design or a progress report to an outside organisation.

1. Visualise yourself through the audience's eyes.
   You want the audience to listen to your message, understand it and be influenced by it. Keep in mind:
   a. The particular concerns of the individuals in the panel (commercial, academic, etc.) are important.
   b. They may not have very much prior knowledge of your work.
   c. What is obvious to you may not be so to them.
   d. The significance to them of each point you make matters, e.g. impact on part numbers, costs and assembly time without reducing the quality of the product.
   e. Summarise the take-home message. It can be couched in terms of economic feasibility, fixed and variable cost savings, projected break-even points, payback period, etc.
   f. The possible barriers to getting your ideas accepted need to be identified.
   g. Concrete examples, not concepts, are preferred by the majority of the population.

2. Identify the key points. Then prepare a short presentation either by computer/projector or by hard copy.
   a. **Identify whether you need a projector.** If there is not one in the room, ask whether you can take one. If there is no screen, establish whether you can use the wall to project onto.
   b. **Prepare your presentation also as a handout.** Make enough copies for all the people on the panel with a few to spare.
   c. **Be rigorously selective in what you will present.** There is never enough time to say everything.

d. Think graphically: plan the presentation around your illustrations. No review panel will react well to screenfuls of dense text. Scientists and engineers think graphically; they are much happier with schematics, graphs, illustrations, etc.

e. Make sure that the quality of the graphical presentation is excellent.

f. Aim to present all your main material in the first few minutes. Use the same diamond structure as for a conference presentation (see Figure 19.1, Chapter 19: *A Seminar or Conference Presentation*, page 236).

- An initial overview of the main points. Make sure you first present – *very briefly* – the context (background) of the work and the reasons you are doing it (the gap in the knowledge, the motivation).
- The main body of the work.
- The main conclusions.

g. Identify the main points (the take-home messages). Make sure that you clearly transmit them, both in the presentation and in the discussion.

h. In addition to the obvious points of your work in the presentation and handout, make sure that you also include the following:

- **An initial overview slide** containing the main points of your work.
- **A graphic that summarises your approach to the project.**
- **Any key recommendations.** Don't overwhelm people with a large number of recommendations – prioritise, stating the important features.
- **A slide showing the current status of the project**, so that it can be presented at a moment's notice.

Checklist for a Presentation to a Small Panel

- ☐ Visualise yourself and your material through the audience's eyes.
- ☐ Work out beforehand the questions you may be asked.
- ☐ Identify the main points of the work and its strengths.
- ☐ Identify the key weak points and problems, and prepare yourself for questions about them.
- ☐ Think graphically. Clear graphical visual aids are an effective means of making points and answering questions.

For a Ph.D. oral examination

- ☐ Have you allowed yourself adequate time to become thoroughly familiar with the material and the thesis again?
- ☐ Have you prepared a summary of the work that you've done?
- ☐ Have you thought about the significance of your work: what you've done, how you've done it and what's new about your research?
- ☐ Are you sure you can navigate your way around your thesis without hesitation?
- ☐ Have you recently done a literature search for any new work that may have come out since submitting your thesis?

- ☐ Can you answer the following questions:
  What is the significance of your work?
  What skills did you develop?
  If you were to do it again, would you approach it differently?
  Where do you see it leading?
- ☐ Have you worked out other questions you might be asked?
- ☐ Can you identify the main points of the work and its strengths?
- ☐ Do you know its weak points, and are you prepared for questions on them?

For a review panel

- ☐ Do you know the particular concerns of the individuals in the panel (commercial, academic, etc.)?
- ☐ Do you know how much prior knowledge of your work they have?
- ☐ Can you gauge the significance to them of each point you make?
- ☐ Can you summarise the take-home message of your work?
- ☐ Can you couch it in terms that are meaningful to the panel (economic feasibility, fixed and variable cost savings, projected break-even points, payback period, etc.)?
- ☐ Can you identify any possible barriers to getting your ideas accepted?
- ☐ Have you been rigorously selective in what you will present?
- ☐ Can you present all your main material in the first few minutes?
- ☐ Have you prepared an initial overview of the main points and a final summing up?
- ☐ Will you present the take-home message – the main point – three times: initial overview, the main body and final summing up?
- ☐ Have you included the following information?
  - ☐ An initial overview slide containing the main points
  - ☐ A graphic that summarises your approach to the project
  - ☐ Clearly demonstrated key recommendations
  - ☐ An ongoing presentation that shows the current status of the project
- ☐ Are your graphics excellent?

Appendices

Appendix 1 SI Units and Their Abbreviations

For greater detail, use the authoritative online source: *National Institute of Standards and Technology. International System of Units (SI):* http://physics.nist.gov/cuu/Units/units.html

SI Base Units and Symbols

Quantity	Name	Symbol
Base Units		
Length	metre	m
Mass	kilogram	kg
Commonly used unit of mass	gram	g
Time	second	s
Electric current	ampere	A
Temperature	degrees Kelvin	°K
	degrees Celsius (acceptable for experimental temperature)	°C
Volume	cubic metre	m^3
Commonly used unit of volume	cubic centimetre	cm^3
Amount of substance	mole	mol
Luminous intensity		cd
Supplementary units		
Plane angle		rad
Solid angle		sr

Other Units Used with SI

Name	In Terms of Other Units	Symbol
atmosphere	101,325 Pa	atm
calorie	4.18 J	cal
day	24 h	d
degree	(π/180) rad	°
hour	60 min	h
kilogram-force	9.8067 N	kgf

Name	In Terms of Other Units	Symbol
litre	1 dm^3	l
micron	10^{-6} m	π
minute	60 s	min
minute	(π/10,800) rad	'
angular second	(π/648,000) rad	"
tonne	10^3 kg	t
torr	133.322 Pa	torr

Examples of SI Derived Units

Quantity	Name	Symbol	In Terms of Other Units
Activity of a radionuclide	becquerel	Bq	s^{-1}
Acceleration			m/s^2
Capacitance	farad	F	C/V
Current density			A/m^2
Electric charge, quantity of electricity	coulomb	C	sA
Electric potential, electromotive force, potential difference	volt	V	W/A
Energy, work, quantity of heat	joule	J	N m
Energy density			J/m^3
Force	newton	N	$(m\ kg)/s^2$
Frequency	hertz	Hz	s^{-2}
Heat capacity, entropy			J/K
Illuminance	lux	lx	lm/m^2
Luminance			cd/m^2
Luminous flux	lumen	lm	cd sr
Magnetic flux	weber	Wb	V s
Moment of force			N m
Power, radiant flux	watt	W	J/s
Pressure, stress	pascal	Pa	N/m^2

Standard Prefixes Used with SI Units

A prefix is a verbal element used before a word to qualify its meaning, e.g. *milli*metre (mm) – a thousandth of a metre; *kilo*metre (km) – a thousand metres; *milli*litre (ml) – a thousandth of a litre; etc.

Term	Multiple	Prefix	Symbol
10^{24}	1 000 000 000 000 000 000 000 000	yotta	Y
10^{21}	1 000 000 000 000 000 000 000	zetta	Z
10^{18}	1 000 000 000 000 000 000	exa	E
10^{15}	1 000 000 000 000 000	peta	P
10^{12}	1 000 000 000 000	tera	T
10^{9}	1 000 000 000	giga	G
10^{6}	1 000 000	mega	M
10^{3}	1 000	kilo	k
10^{2}	1 00	hecto	h
10^{1}	10	deca	da
	1 unit		
10^{-1}	0.1	deci	d
10^{-2}	0.01	centi	c
10^{-3}	0.001	milli	m
10^{-6}	0.000 001	micro	μ
10^{-9}	0.000 000 001	nano	n
10^{-12}	0.000 000 000 001	pico	p
10^{-15}	0.000 000 000 000 001	femto	f
10^{-18}	0.000 000 000 000 000 001	atto	a
10^{-21}	0.000 000 000 000 000 000 001	zepto	z
10^{-24}	0.000 000 000 000 000 000 000 001	yocto	y

Appendix 2 The Parts of Speech; Tenses and Forms of the Verb

Parts of Speech

Parts of Speech	The Work That Words Do in a Sentence
Verbs	Words that indicate action: what is done, or what was done, or what is said to be. The ship *sailed.*
Nouns	Names. Things. *Columbus* sailed in the *ship.*
Pronouns	Words used instead of nouns so that nouns need not be repeated. *He* sailed in *it.*
Adjectives	Words that describe or qualify nouns. The *tall* man sailed in the *big* ship.
Adverbs	Words that modify verbs, adjectives and other adverbs. They often end in *-ly.* The big ship *slowly* sailed past the *steeply* sloping cliffs.
Prepositions	Each preposition marks the relation between a noun or pronoun and some other word in the sentence. The ship sailed *past* the cliffs and *across* the sea *to* America.
Conjunctions	Words used to join the parts of a sentence, or make two sentences into one: **and, but, so, because, as, since, while**. The ship sailed to America *and* came straight back. The ship sailed to America *but* did not stay long. The ship sailed fast, *so* it got there quickly. The ship sailed slowly *because* (or *as* or *since*) the sails were torn. The people on the dock waved *while* the ship sailed away.
Gerund	A word ending in *-ing* that behaves in some ways like a noun and in some ways like a verb. She likes *using* a computer You can save electricity by *switching* off the lights.

Tenses and Forms of the Verb

This section describes, in very simple terms, the various forms of a verb.

For guidelines in their use in various sections and documents, see **The Correct Form of the Verb**, Chapter 18: *Problems of Style*, page 224.

Present

Describes what is happening at the moment:

1. The sun *shines.*
   Chlorofluorocarbons *cause* ozone depletion.
2. The sun *is shining.*
   Climate change *is developing* into a major global issue.

Past

Describes what happened in the past.

1. The sun *shone.*
   The burning of fossil fuels *caused* carbon dioxide levels to rise.
2. The sun *was shining.*
   By the end of the twentieth century, carbon dioxide levels *were causing* temperature levels to rise.
3. The sun *has shone.*
   The burning of fossil fuels *has caused* levels of carbon dioxide to rise.
4. The sun *has been shining.*
   Carbon dioxide levels *have been causing* concern for a long time.
5. The sun *had shone.*
   By the 1950s, carbon dioxide levels in the atmosphere *had risen* to 315 ppm.
6. The sun *had been shining.*
   By the end of the 20th century, carbon dioxide levels *had been rising* for a number of decades.

Future

Describes what is going to happen in the future.

1. The sun *will shine.*
   Increased emission of greenhouse gases *will cause* a change in the global climate.
2. The sun *will be shining.*
   By the middle of this century, the increased emission of greenhouse gases *will be causing* a global change in climate.
3. The sun *will have shone.*
   By the middle of this century, carbon dioxide levels *will have risen* to twice the pre-industrial level.
4. The sun *will have been shining.*
   By the middle of this century, carbon dioxide levels *will have been causing* concern for many decades.

Conditional

These express a condition and are sometimes needed in recommendations:

This result could imply that …
The test equipment should be modified as shown.

Subjunctive

In technical writing, these are usually used only in recommendations.

It is recommended that the system be upgraded.
It is recommended that the manager assess the effects of the change.

Imperative

This form of the verb gives an instruction. This is the preferred form of the verb for a procedure or set of instructions.

Turn the power off.
Do not open Valve X before it cools to 18°C.

Appendix 3 Recommended Scientific Style Manuals

The following two sources are recommended for authoritative information on all aspects of style, regardless of your subject area.

One of the most used manuals in scientific style and format:

For All Areas of Science and Related Fields

Scientific Style and Format: The CSE Manual for Authors, Editors and Publishers. 7th edition. Council of Science Editors, 2006. Hard copy only.

The CSE Manual is the most recognized, authoritative reference for authors, editors, publishers, students and translators in all areas of science and related fields. The manual is usually known by its subtitle and is highly recommended.

For General Style

The Chicago Manual of Style: For Authors, Editors and Copywriters. Chicago: University of Chicago Press. Online: http://www.chicagomanualofstyle.org/home.html

Printed in the United States
By Bookmasters